Space and Society

Series editor

Douglas A. Vakoch, Mountain View, CA, USA

More information about this series at http://www.springer.com/series/11929

Urania and the celestial sphere. Marble, I century CE, from Villa Adriana, Tivoli. Pio Clementino Museum, Vatican Museums. Photo by Jastrow—Wikimedia Commons

Leonardo Magini

Stars, Myths and Rituals in Etruscan Rome

Foreword by Germaine Aujac

Springer

Leonardo Magini
Società Italiana di Archeoastronomia
Fiumicino
Italy

Translation by Adam Victor

ISSN 2199-3882 ISSN 2199-3890 (electronic)
ISBN 978-3-319-36001-0 ISBN 978-3-319-07266-1 (eBook)
DOI 10.1007/978-3-319-07266-1

Springer Cham Heidelberg New York Dordrecht London

Printed on acid-free paper

Springer is part of Springer Science+Business Media (www.springer.com)

For Alessandro

Nessuna umana investigazione
si puo' dimandare vera scienza,
s'essa non passa per le matematiche
dimostrazioni.
(No human investigation can be called
true science without going through
the mathematical tests.)

Leonardo da Vinci

Foreword

Etrusco-Roman Astronomy is a brilliant summation of Leonardo Magini's prior works. Since his earliest endeavours (*Le Feste di Venere*, 1996, *Astronomy and Calendar in Ancient Rome*, 2001), not to mention his works on the Etruscan language (*La Parola degli Etruschi*, 1987 and *L'Etrusco, lingua dall'Oriente Indoeuropeo*, 2007), all of his research has amply demonstrated the author's interest in the origins of Roman civilization, and just how much Roman civilization owes to the culture of its Etruscan neighbours.

Magini draws on an impressive number of Latin and Greek texts which, though indubitably hailing from later periods, nevertheless make up a well-consolidated body of opinion. He collates astronomical notions and data on religious feast days to highlight the importance of the revised calendar traditionally ascribed to the legendary king Numa Pompilius, who is generally assumed to have been the heir to the founder of Rome. Whereas the year in use under Romulus had but 10 lunar months—a characteristic that soon put it out of kilter with the Sun's motion—Numa corrected this error not just by adding a further two lunar months, but by establishing a cycle of three times 8 years which, using different numbers of intercalated days, almost wholly restored the year's concordance with the solar year.

The cycle attributed to Numa is an extremely close fit to the motion of the major objects in the sky—the Sun, the Moon and Venus—and could be used to predict eclipses of the Sun and Moon. In other words, it was a worthy successor to the cycle of Saros, a fact that itself points to a Mesopotamian influence. Did not the Greek historian Herodotus say that the Etruscans were descended from immigrants who originated in Lydia?

Even relative neophytes to astronomical issues will be fascinated by the relationship that the author establishes and proves between the passage of the year as marked by the Numan cycle, and the cadence of the many religious celebrations that set apart the *fasti* (festive days, and the only days when the courts were in session in Ancient Rome) from normal days (when people had to work, and justice could not be dispensed). Every feast day offers up an "explanation" to the passage of the heavenly bodies.

Among the merits of this groundbreaking work is its clear demonstration of the extraordinary astronomical knowledge contained within the Roman religious calendar, which was established at Rome's very beginnings. Though by no means certain, it is certainly legitimate to presume that this knowledge was strongly influenced by Etruscan knowledge, and, before them, by the science of the Chaldeans. The author's lines of enquiry should be considered as an initial proof, to be confirmed by additional research.

Germaine Aujac

Preface: A Letter

To my friends, Professors Franco Aspesi, Gioachino Chiarini and Mario Negri[1]

Distinguished professors and dear friends,

As you have followed my research into the astronomical underpinnings of the Numan calendar right from the start, I send you a monograph that I have written to address multiple objectives.

First of all, the essay brings together the fruits of several years' study, as illustrated in my books *Le Feste di Venere* and *Astronomy and Calendar in Ancient Rome*,[2] supported by the indispensable writings of authors of antiquity and astronomical calculations. This essay also draws on fresh calculations and analyses to build on my previous work and illustrates more recent developments, including a new approach here and there. Last but not least, the essay opens up a number of avenues for further research which—I am the first to admit—a sense of satiety with the work I have undertaken thus far leads me to believe will be for someone else to pursue. I am only too happy to sketch out these avenues so that fresh blood may consolidate and advance research into the now burgeoning discipline of Etrusco-Roman astronomy.

I would, however, like to take this opportunity to make a few remarks regarding the point at which, I believe, we have arrived.

First, the sum of astronomical and astrological knowledge held by the anonymous inventor of the Numan calendar is codified in the temporal rhythms of the year's festivities, fitted into an appropriate system of intercalation.

Second, the rhythm of these cadences, associated with reading and interpreting the myths and rites associated with the festivities, allows us to track backwards to real astral movements.

---

[1] This letter, written in 2003 for the first edition of *Astronomia etrusco-romana* (*Etrusco-Roman astronomy*), was also sent to Vittorio Castellani, who passed away in 2006.

[2] See, respectively, Magini 1996 and Magini 2001.

Third, the astronomical knowledge thus codified includes: the motion of Venus; the periodicity of solar and lunar eclipses; the revolution of the line of lunar nodes; the passage of nodes from the Points of Aries and Libra, and the effect that this has on the motion of the Moon; the revolution of the line of apsides; and the relation between the synodical and sidereal periods of the superior planets. I touch only briefly on areas where certainty remains elusive: the rhythm of the motion of the superior planets, particularly Jupiter and Saturn, and the position of zero degrees Aries. Lastly, work still needs to be done to clarify the astronomical significance of the nundinal letters, the true significance of the days of Agonalia (March 17, May 21, December 11 and January 9), and of the opening of the *mundus* (August 24, October 5 and November 8).

Fourth, we know without a shadow of doubt that there was an astrological tradition in Rome, among other things, because of the protection afforded to women—the guarantors of the continuity of the species—by the two most "feminine" of celestial bodies since the proto-history of astrology: the Moon and Venus. This tradition is further evident in the bonds between the public exchequer and the Temple of Saturn from 497 BCE (at the very latest) onwards.

Fifth, last, and most important of all, the very rhythm of public and private life during the time of the Roman Kingdom was regulated by the motion of celestial bodies. Public life was ruled by the *Vestalia,* the *Regifugium* and the *interrex*, and by the *October Equus,* the *Fordicidia* and the *Parilia.*[3] Private life was wholly ruled by the celestial bodies. Conception, the seventh month of pregnancy, preparation for birth, birth itself, namegiving, all of the rites of passage associated with puberty, marriage and conception anew, all of which are notable life landmarks—particularly in the lives of women—were cadenced by celestial movements and the rhythm of their associated festivities: *Matralia* and *Carmentalia, Matronalia* and *Liberalia*, *Tigillum sororium* and *Anna Perenna, Veneralia* and, once again, *Matralia.*[4] Indeed, the circularity of human life on Earth reflects the circularity of the motion of the celestial bodies in the heavens.

Having established these foundations, our enquiry must shift to their origins. An initial hypothesis is that these foundations descend directly from the prehistory of the Roman people. Such a hypothesis, however, flies in the face of the entire weight of tradition, which has it that King Numa was responsible for establishing the calendar, intercalation,[5] and a great many of the feast days. There is also the matter of explaining why no other ancient Italian population, despite significant similarities in many respects, was able during the same period of time to achieve a similar level of astronomical and calendar knowledge. Indeed, how could it have been possible in historical times for the Romans to "misplace" the origins of a

---

[3] See Magini 2001, pp. 59–67 and 93–104; these topics are covered, respectively, in Chaps. 16 and 17.

[4] See Magini 1996, *passim*; in this book, the topic is covered in Chaps. 10 and 11.

[5] Brought in *sacrorum causa*, "for religious feast days"; see Valerius Antias, fr. 5 Peter apud Macrobius, *Saturnalia*, 1.13.20.

cultural and intellectual heritage acquired through endless years of research and complex mental processes, ever since prehistoric times?

An alternative hypothesis would lead us to a source from outside the Roman sphere. Possible solutions may be found in similarities between the *Regifugium-interrex-Vestalia* and the *sar puhi,* the "substitution of the king" ritual from Ancient Babylon, where the same landmarks in astronomical knowledge were attained a number of centuries earlier. If this association proves to be correct, the question arises as to how this knowledge and its vast store of rituals and conceptions managed to reach Rome, leapfrogging the Greek world.

It is my personal belief that the answer lies elsewhere. I believe that the answer is to be found in the Etruscan origin of the names of the great many festivities associated with astronomical timings. The Etruscans had the *Nonae*, the name *Ides,* the *nundinae,* the *clavus annalis*, and—this is the clincher—the Capitoline triad of Jupiter, Juno and Minerva. This is where we should seek the "barbarian superior to Pythagoras" who, according to Plutarch, put the finishing touches to educating the Roman king. And yet this adds a further layer of complexity, for Pythagoras himself is complex enough: known as "Tyrrhenian" by birth and upbringing, Pythagoras studied geometry, mathematics and astronomy in Egypt and Mesopotamia, before reputedly teaching so much to Numa—though not, it turns out, the calendar to which the king would lend his name.[6]

The picture only becomes much clearer if we accept that the Etruscans hailed from the Orient—from Lydia, as Herodotus claims. This would appear to be confirmed by "the sale of Sardians", that is to say, the Etruscans whom Romulus captured at Veio.[7] First, the Mesopotamian/Anatolian area had achieved a whole sequence of scientific achievements and an elaborate worldview that was the culmination of at least two millennia of development. Thereafter, over what can in no way be described as a brief period of time, an educated and cultured élite belonging to a population living on the fringes of this area acquired these notions of astronomy and astrology, along with a worldview that gave them a rounded meaning. Finally, as events unfolded, this population—including its élite—travelled to Italy, defeated Rome, and as conquerors imposed a calendar that, far from being "agricultural/pastoral" as has long been claimed, was based on a surprisingly sophisticated accretion of astronomical

[6] On the barbarian, see Plutarch, *Life of Numa*, 1.
On the birth and upbringing of Pythagoras, see Aristoxenus of Tarentum, fr. 1, in F. H. G., p. 272; Diogenes Laertius, 8.1; Neanthes of Cyzicus, fr. 30, in F. G. H. 3, p. 10 (the Tyrrhenians/Pelasgians of Samos and Lemnos, to whom these three authors refer, are the Eastern branch of the Tyrrhenians/Etruscans; these branches were united not just by language but by their iron technology).
On his education, Plutarch, *Conviviales Quaestiones* 8.727B; on P.'s knowledge, Giamblicus, *Vita Pytagorae*; on relations between P. and Numa, Plutarch, *Numa*, passim.

[7] Herodotus, *Histories*, 1.94.
On the "Sardians for sale", Plutarch, *Romanae Quaestiones*, 53; Sardis was the capital of the kingdom of Lydia.

knowledge. This knowledge was initially held by the élite; given its complexity, it was impossible to pass it down through the ensuing centuries.

Pursuit of this line of enquiry—which today would be validated by examination of "the genetic data of the populations in potential areas of origin", to quote Cavalli-Sforza[8]—would also explain the sudden step change in the Roman calendar, from the mysterious, cobbled-together 304-day Romulean Calendar (unless behind this astronomically meaningless number lies a 61-day interval between significant astronomical phenomena, but then what phenomena would they be?)[9] to the complex and sophisticated calendar adopted by the following generation (i.e. Numa's generation), which was light years ahead of the preceding simple—and basic—lunar calendar.

I believe that the same path was followed by other customs, beliefs and ideas regarding the relationship between men and divinities or, more specifically, the life of man and celestial messages. I refer to haruspicy: divination through the examination of sheep livers, the various parts of which represented parts of the heavens ruled by the planets and constellations, each of which in turn represented a divine being; I refer to brontoscopic calendars, in which every day of the year corresponded to a particular sign belonging to a particular god; and I refer to the "the curse of the number nineteen". Indeed, in Ancient Babylon, where every single month was 30 days long, the number 49 was bad luck: "The week of weeks was determined by the 49th day after every new Moon—in a 30-day month, that meant that it ran from the 19th day of the following month. Hemerologists considered this nineteenth day to be an *umu limnu,* an inauspicious day… The Babylonians avoided the ill-omened 19th, and even went so far as to write the number as XX.I.lal, i.e. 20 *minus* 1."[10]

This credence may have resurfaced in Etruria;[11] I say "may have"—with the stress on its uncertainty—because as with almost everything to do with Etruria, studies in the field, most especially in Italy, have suffered from the calamitous impact that ideology is capable of having on science. The credence, however, certainly does reappear in Ancient Rome, in the form of *undeviginti,* literally "one-away-from-twenty", despite the fact that Rome had no need for such counting tricks, given that it had no residual trace either of the concepts of a "week of weeks", nor of 30-day months…

Above all, I believe it impossible to attribute such a path of transmission—from the East to Etruria and on to Rome—to a fashion for "orientalizing". Fashion and

---

8 Cavalli Sforza 2000, p. 705.

9 A putative answer is presented in Chaps. 26 and 27.

10 On haruspicy, see Nougayrol 1955, pp. 509 and following; also, see Nougayrol 1982, p. 9, no. 7, and pp. 104 and 107; Bottéro 1982, pp. 73–214.
On brontoscopic calendars, see Joannes Lydus, *de ostentis*, passim; Bottéro 1982, pp. 108–11; Pettinato 1998, pp. 196–8.
On the "curse of the number 19", see Schiaparelli 1997–1998, Vol. I, p. 299.

11 Caffarello 1975, p. 110; with reference to Trombetti, Pallottino and Pfiffig, though in such cases *repetita* not *iuvant.*

Fig. 1 "Orientalizing" patera. Gilded silver, 19.4 cm in diameter, seventh century BCE, Cerveteri, Regolini-Galassi tomb—Vatican Museum, Rome. Courtesy Vatican Museum

bragging have little use for Calendars; Calendars are not Assyrian paterae (Fig. 1) or Urartean cauldrons, as is evident from two thousand years of European calendar history. Fashion cannot explain why people have studied, observed and recorded the movements and rhythms of the stars to try and understand future events; why the planet Venus was considered to be the representative of a feminine divinity of evening love—Ishtar in Mesopotamia and Venus in Rome, and a divine protector of morning warfare—Ishtar once again, plus Mater Matuta; and, with regard to celestial bodies in general—fixed stars and planets, the Moon and the Sun—considering them as heaven's corollaries to the gods, an expression of their desires which it is our lot to interpret.

It could hardly have been turbocharged Villanovans (turbocharged by what, pray tell?)—and I say this in open dispute with the triumphant curators of exhibitions in Bologna and Venice[12]—who came up with the idea of a direct and binding relationship between the macro-cosmos and the micro-cosmos, between

[12] *Principi etruschi tra Mediterraneo ed Europa*, conceived and curated by G. Bartoloni, F. Delpino, C. Morigi Govi, and G. Sassatelli, Museo Civico Archeologico, Bologna, October 2000—April 2001. *Gli Etruschi*, curated by M. Torelli, Palazzo Grassi, Venice, November 2000—April 2001.

the motion of the celestial bodies across the sky and the life of men on this Earth. If, as Seneca writes in historical times,

> This is the difference between us Romans and the Etruscans […]: We believe that lightning is caused by clouds colliding, whereas they believe that clouds collide in order to create lightning. Since they attribute everything to gods, they are led to believe not that events have a meaning because they have happened, but that they happen in order to express a meaning,[13]

it is clear that these approaches are so very distant that not even the preceding millennium, with its mingling of Roman and Etruscan peoples and cultures, was able to generate a common conception of the world.

Where we may find such a common conception, however—in my opinion, though perhaps not in yours—is in the way that Seneca depicts the Etruscans, and the way that the Babylonians are portrayed in the "Diviners' Manual":

(23) […] their good and their evil portents are in harmony (i.e. confirming each other).
(24) the signs in the sky just as those on the Earth give us signals […]
(37) […] their good and evil portents
(38) are in harmony. The signs on Earth just as those in the sky give us signals.
(39) sky and Earth both produce portents:
(40) though appearing separately, they are not separate (because) sky and Earth are related.
(41) a sign that portends evil in the sky is (also) evil on Earth,
(42) one that portends evil on Earth is evil in the sky […]
(53) […] in summa 25 tablets with signs (occurring) in the sky and on Earth
(54) whose good and evil portents are in harmony (?)
(55) you will find in them every sign that has occurred in the sky
(56) (and) has been observed on Earth […][14]

There is no need for me to repeat here words pronounced by others—far more eloquently than I could possibly manage—regarding the pre-scientific value (in the sense of preparatory to a scientific approach) of a mental outlook of this nature: Pannekoek (a historian of science) and Bottéro[15] (a historian of civilization) have expounded upon this in exemplary and illuminating fashion.

What I would like to do, however, is offer some observations on Bottéro's comment: "The Greeks were not born into a world of Primates, in a kind of barren, cultural vacuum; borrowing an expression uttered by an Ancient Greek on a

[13] Seneca, *Naturales Quaestiones*, 2.32.2: *Hoc inter nos et Tuscos […] interest: nos putamus, quia nubes collisae sunt, fulmina emitti; ipsi existimant nubes collidi, ut fulmina emittantur. Nam cum omnia ad deum referant, in ea opinione sunt tanquam non, quia facta sunt, significent, sed quia significatura sunt, fiant.*

[14] Oppenheim 1974, pp. 197–211.

[15] Pannekoek 1989, pp. 36–81; Bottéro 1982, pp. 187–211.

different topic, what they did was 'they transformed' what they learned 'into something more beautiful'. The things that they thus transformed were 'borrowed from the Barbarians', especially, notes the author, from the Ancient Mesopotamians *(Epinomides, 987e-988a and 987a).*"

I believe that, similarly, the Etruscans were not born into a world of Primates; a world in which the only civilized people were rich but uncultured gentlemen from the East, from whom they could import luxury items, while from the poor but wise Greeks they were able to plunder—more than learn—everything else. The Etruscans too had their sum of knowledge, their own *weltanschauung,* which they attempted to bequeath to the Romans, albeit with limited success. Much of what we are able to reconstruct, such as the Etruscan calendar, survived its creators but in the process lost almost all of its meaning for the people who continued to use it. If my reconstruction is correct, the label of "Roman" rather than "Etrusco-Roman" astronomy would be an umpteenth abuse of power, of history being written by the winner.

I fear that I may have strayed from my own sphere of competence; I shall rein myself back in. Indeed, I wish to take up the thread from where and how I began, from the "planned coincidence" that first led me to undertake a whole series of minor calculations on the calendar, and to note that in the Numan calendar 71 days elapse between the *Veneralia* and the *Matralia.* This process led me to read (frankly absurd) books such as Graves' *The White Goddess*. I may never have finished reading that book, but I recall perfectly the day when I bought it, opened it, turned to the index, and looked under Venus to find Venus (planet). The first reference, on page 293 of my Italian edition, says: "72 [...] is also linked to the Goddess astronomically, by the 72-day season during which her planet Venus moves successively from maximum eastern elongation to inferior conjunction (its closest approach to Earth) and thence to maximum west elongation."[16] This was what first triggered my research 7 or 8 years ago. Ever since then, I have been unsure quite where my sphere of competence begins and ends.

One thing I do know, however, is that I haven't invented a bean of all this. All of my reconstructions are based on well-known, if not to say renowned, documents. I have based my theses on nothing new—nothing that hasn't already been known and been studied for at least two thousand years. I believe that I am the first person—to my knowledge, at least—to prove the existence of a wealth of scientific knowledge in the classical sources I have used and interpreted, merely by investigating the anthropological, mythical and religious data that they contain. All of this was revealed through an investigation of the astronomical underpinnings of the Numan calendar.

One further thought springs to mind. Why was it that Mesopotamian astronomy was "discovered" a century or more before Roman astronomy? Was it simply because of the long-held credence that the Romans knew nothing of astronomy? Can conventional thinking really have had such a negative influence?

[16] Graves 1992.

Mesopotamian astronomy was "discovered" because scholars were open-minded, notwithstanding only minimal previous evidence of the Chaldeans' astronomical knowledge. We may legitimately wonder how open-minded scholars have been regarding Ancient Rome, and how much they have been influenced—as modern-day scholars of the Etruscans continue to be, at least in Italy—by a preconceived ideological approach.

It remains for me to thank you warmly for taking the time to read this work, in the hope that it proves to be interesting and enlightening.

Torrimpietra, March 2002–2014 Leonardo Magini

Contents

Figures

Tables

Notes

Chaps. 1 and 21 have been excerpted from the paper *Astronomia e Calendario nell'antica Roma*, given at the First Conference of the Italian Archeo-astronomy Society, in Padua, on 29 September 2001 (see Rivista Italiana di Archeoastronomia, vol. I, 2003, pp. 107–19).

Chapter 3 is from the paper *L'etrusco, lingua dell'Oriente Indoeuropeo I*, given at the Milan Glottological Fellowship on 14 June 1999 (see Atti del Sodalizio Glottologico Milanese, Vol. XXXIX–XL, 1998–9, pp. 229–49; reprinted in Magini 2007, pp. 21–7).

Chapters 4, 13–18 are extracted from *Astronomy and Calendar in Ancient Rome—The Eclipse Festivals*, Rome 2001; Chaps. 4, 6, 13, 14, 17, 9 and 10 have previously appeared in the paper *Cicli astronomici e feste del calendario di Roma antica*, given at the Conference on the History of Astronomy at the Capodimonte-Naples Observatory on 27 September 1997.

Chapters 7–11 are extracted from *Le feste di Venere—Fertilità femminile e configurazioni astrali nel calendario di Roma antica*, Rome 1996, and from *I moti di Venere e le feste delle donne nel calendario di Roma antica*, given at the Fourth Conference of the Italian Archeoastronomy Society, in Lerici on 24–25 settembre 2004 (see "Rivista Italiana di Archeoastronomia", vol. IV, 2006, pp. 123–34).

Chapters 8 and 23 are from the paper *L'etrusco, lingua dell'Oriente Indoeuropeo IV: nomi di dèi e moti di astri*, given at the Milan Glottological Fellowship on 4 March 2002 (see Atti del Sodalizio Glottologico Milanese, Vol. XLIII–XLIV, 2002–3; pp. 49–62).

Chapter 12 is taken from the paper *Le feste dei solstizi in Roma antica*, given at the Seventh Conference of the Italian Archeoastronomy Society, in Rome, october 2007, being printed; see also *La dea bendata—lo sciamanesimo nell'antica Roma*, 2008, pp. 207–17.

Chapter 16 is taken from the paper *Eclissi e regalità: un rapporto difficile (Babilonia 1.900 a.C—Roma 1.600 d.C.)*, given at the Third Conference of the Italian Archeo-astronomy Society, Capodimonte Observatory, Naples, 26–27 settembre 2003, in "Rivista Italiana di Archeoastronomia", III. 2005, pp. 1–5.

Chapter 26 is taken from the paper *Il calendario romuleo e i suoi rapporti con i fenomeni astronomici*, given at the Second Conference of the Italian Archeoastronomy Society, Monte Porzio Catone, on 28 September 2002 (see Atti del II Congresso di Archeoastronomia, vol. II, 2004, pp. 77–81); see also *The astronomical foundations of the Romulean calendar and its relationship with the Numan calendar: an hypothesis,* in *Automata—Rivista di Natura, Scienza e Tecnica del mondo antico*, n. 3, « l'Erma » di Bretschneider Editore, Roma 2009.

Chapter 27 is taken from the paper *Il passaggio dall'anno romuleo all'anno numano e lo slittamento del solstizio d'inverno* given at the Tenth Conference of the Italian Archeo-astronomy Society, Trinitapoli, 22–23 ottobre 2010, being printed.

Other topics in this book have previously been covered in my seminar *I fondamenti astronomici del calendario numano:* Anna Perenna, October Equus, Regifugium *e* Vestalia, given at the University of Siena on 22 March 1999, and at lessons held at the Roma Tre University on 7 May 1998, at the Milan University Institute of Modern Languages—IULM, on 4 March 2002, and at the University of Siena on 17–18 April 2002.

Parameters Adopted for Calculations

I have adopted the following parameters in my calculations:

Sidereal year	365.2422 average solar days
Sidereal month	27.3216 average solar days
Synodical month	29.5306 average solar days
Draconitic month	27.2122 average solar days
Anomalistic month	27.5546 average solar days
Angular velocity of the Sun	0.9856° per day
Angular velocity of the Moon	13.1764° per day
The Moon's variance from the Sun	12.1908° per day
Retrograde revolution of lunar nodes with respect to the fixed stars	6,793.48 days, equivalent to − 0.05299° per day
The Sun passes through a lunar node every	173.31 days
Direct revolution of the apsides with respect to the fixed stars	3,232.61 days, equivalent to + 0.11135° per day

Chapter 1
The Beginning of Astronomical Time and the *Feriae Martis*

The third book of Ovid's *Fasti* is dedicated to March, the first month of the Numan year. Right at the very beginning, before commencing his description of the month's feasts, rites and myths, the poet writes of Romulus's conception:

> Silvia the Vestal (for why not start from her?)
> went in the morning to fetch water to wash the holy things.
> When she had come to where the path ran gently down the sloping bank,
> she set down her earthenware pitcher from her head.
> Weary, she sat her on the ground and opened her bosom
> to catch the breezes, and composed her ruffled hair.
> While she sat, the shady willows and the tuneful birds
> and the soft murmur of the water induced to sleep.
> Sweet slumber overpowered and crept stealthily over her eyes,
> and her languid hand dropped from her chin.
> Mars saw her; the sight inspired him with desire,
> and his desire was followed by possession,
> but by his power divine he hid his stolen joys.
> Sleep left her; she lies big, for already
> within her womb there was Rome's founder.[1]

Thus was Romulus conceived (Fig. 1.1), during what is, as we can surmise from the Vestal's sleep, an undefined moment in time. The poet fails to mention the other event—the other conjunction—that on high accompanied the carnal

[1] Ovid, *Fasti*, 3.1–166; translation by J.G. Frazer. Quoted here are lines 11–24: *Silvia Vestalis—quid enim vetat inde moveri?— / sacra lavaturas mane petebat aquas. / Ventum erat ad molli declivem tramite ripam: / ponitur e summa fictilis urna coma; / fessa resedit humo ventosque accepit aperto / pectore, turbatas restituique comas. / Dum sedet, umbrosae salices volucresque canorae / fecerunt somnos et leve murmur aquae. / Blanda quies furtim victis obrepsit ocellis / et cadit a mento languida facta manus. / Mars videt hanc visamque cupit potiturque cupita / et sua divina arte fefellit ope. / Somnus abit, iacet ipsa gravis. Iam scilicet intra / viscera romanae conditor Urbis erat.*

L. Magini, *Stars, Myths and Rituals in Etruscan Rome*, Space and Society,
DOI 10.1007/978-3-319-07266-1_1

Fig. 1.1 Mars and Rhea Silvia. Romulus was conceived when the god of war came down to Earth and conjoined with the sleeping vestal virgin; legend has it that the carnal conjunction occurred at the same time as a total solar eclipse. Sarcophagus Mattei detail, 220 CE, Palazzo Mattei and Vatican Museums, Rome. Photo by Giovanni dall'Orto—Wikimedia Commons

conjunction on Earth which brought life to the founder of Rome. We rely on the historian Plutarch to discover this, in his essay *On the Fortune of the Romans*:

> […] regarding the generation and conception of Romulus they record that the Sun was eclipsed and came into exact conjunction with the Moon at the time when Mars, a god, consorted with the mortal Silvia.[2]

In his *Life of Romulus*, Plutarch repeats that according to Tarutius' calculations—Tarutius was an astronomer, astrologer and friend of Cicero's—Romulus was conceived by his mother

> during a total eclipse of the Sun.[3]

The start point of astronomical time for the descendants of Romulus begins with the exceptional and simultaneous conjunction of the representatives of the gods—the Sun and Moon in the sky, Mars and Silvia on Earth—with an initial

[2] Plutarch, *de fortuna Romanorum*, 320B: *περι την Ρωμυλου σποραν και καταβολην τον ηλιον εκλιπειν ιστορουσι, ποιησαμενον ατρεκη συνοδον προς σεληνεν, εωσπερ ο Αρης θεος ων τη Σιλβια θνητη συηλθε.*

[3] Plutarch, *Life of Romulus*, 12.5: […] *καθ' ην ο ηλιος εξελιπε παντελως.*

Numan cycle (see Chap. 4) that itself commences with a total eclipse of the Sun on the first day of the first month of year one, celebrated on the feast of *Feriae Martis*.[4]

---

[4] On the relationship between astral phenomena and the life of man, see Proclus (*Comment on Timaeus* 40c; in Garin 1996, p. 132, no. 14): "The occultation and reappearance of heavenly bodies that take place at given moments mark a renewal of the cosmos and the beginning of cycles. Upon these phenomena, earthly things transform and change."

On Romulus, see M. Eliade (*Histoire des religions*, p. 219; cited by Liou-Gille 1980, pp. 162–163), who writes: "The appearance of a given child coincides with an auroral moment: the creation of the cosmos, the creation of new world, a new historical age."

Notably, in the West, our tally of years starts from the birth of a rather special "baby".

Chapter 2
The Numan Year, the Romulean Year and the *Feriale Antiquissimum*

The oldest Western calendar with a clear link to astronomy is traditionally attributed to the second King of Rome, Numa Pompilius (715–673 BCE).

His twelve-month lunar year began in March. Four of his months—i.e. March, May, July and October—were 31 days long; the other months (see Fig. 2.1) lasted 29 days, except for February, which had 28 days (see Summary Chart).

The Numan year was 355 days long, compared with a natural length of 354.3672 ($=12 \times 29.5306$); the absolute error was $+0.6328$ days, or 1.78×10^{-3} in percentage terms.

However, the Roman tradition retained traces of a calendar that predated the "Numan" calendar, which was attributed to the city's founder, Romulus (754–716 BCE),[1] and even to the Albans. The "Romulean" calendar divided the year into 10 months—four 31 day months and six 30 day months—and a total of 304 days. At first sight, there is nothing astronomical about this calendar; however, in later chapters, we will examine the relationship between the Romulean and Numan

[1] On the Romulean and Numan calendar, Censorinus, *de die natali*, 20.3–5, writes: "These 10 months lasted 304 days, distributed thusly: March 31, April 30, May 31, June 30, Quintile (i.e. July) 31, Sextile (i.e. August) and September 30 each, October 31, and November and December 30. The four longer months were known as whole months; the other 6 months were known as incomplete months. Afterwards, Numa […] created 12 months and 355 days […] We know as a certainty that 51 days were added to the old year; as these days were insufficient to form 12 months, 1 day each was removed from the six empty months; these 6 days were added to the 51 days to make a total of 57 days, which were then divided to make 2 months: January, lasting 29 days, and February, lasting 28 […], *Hi decem menses dies CCCIIII hoc modo habebant: Martius XXXI, Aprilis XXX, Maius XXXI, Iunius XXX, Quintilis XXXI, Sextilis et September tricenos, October XXXI, November et December XXX; quorum quattuor maiores pleni, ceteri sex cavi vocabantur. Postea sive a Numa […] XII facti sunt menses et dies CCCLV […] Certe ad annum priorem unus et quinquaginta dies accesserunt; qui quia menses duo non explerent, sex illis cavis mensibus dies sunt singuli detracti et ad eos additi, factique dies LVII, et ex his duo menses, Ianuarius undetriginta dierum, Februarius duodetriginta […]*".

On the Romulean and Numan calendars, see also Macrobius, *Saturnalia*, 1.13.

L. Magini, *Stars, Myths and Rituals in Etruscan Rome*, Space and Society,
DOI 10.1007/978-3-319-07266-1_2

Fig. 2.1 The month of April, detail the second month of the Numan calendar and its festivities. *Fasti Praenestini*, detail. *Fasti praenestini*, between 6 and 9 CE, Palazzo Massimo/Museo Nazionale Romano, Rome. Photo by Marie-Ian Guyen-Wikimedia Commons

calendars and how it may have been possible to move from one to the other (see Chaps. 26 and 27).

Both Romulean and Numan calendars contained feast days. The so-called *Feriale antiquissimum*—a modern name—is a list of officially-sanctioned festivities dating back to the earliest Roman kings. A handful of these feast days—such as *Parilia* and *Saturnalia*—already existed before the foundation of Rome; some of them were established by Romulus; and many others were decreed by Numa Pompilius.

In more recent times, the *Feriale* consisted of forty-five feast days, plus the twelve *Idus* dedicated to Jupiter, making a total of fifty-seven feast days. The following feast days were designated in official calendars using capital letters:

- in March, the *Equirria* on 14, the *Liberalia* on 17, the *Quinquatrus* on 19, and the *Tubilustrium* on 23;
- in April, the *Fordicidia* on 15, the *Cerialia* on 19, the *Parilia* on 21, the *Vinalia* on 23, and the *Robigalia* on 25;
- in May, 3 days of the *Lemuria* on 9, 11 and 13, the *Agonalia* on 21, and the *Tubilustrium* on 23;
- in June, the *Vestalia* on 9, and the *Matralia* on 11;

- in July, the *Poplifugium* on 5, 2 days of *Lucaria* on 19 and 21, the *Neptunalia* on 23, and the *Furrinalia* on 25;
- in August, the *Portunalia* on 17, the *Vinalia* on 19, the *Consualia* on 21, the *Volcanalia* on 23, the *Opiconsivia* on 25, and the *Volturnalia* on 27;
- in October, the *Meditrinalia* on 11, the *Fontinalia* on 13, and the *Armilustrium* on 19;
- in December, the *Agonalia* on 11, the *Consualia* on 15, the *Saturnalia* on 17, the *Opalia* on 19, the *Divalia* on 21, and the *Larentalia* on 23;
- in January, the *Agonalia* on 9, and 2 days of the *Carmentalia* on 11 and 15;
- in February, the *Lupercalia* on 15, the *Quirinalia* on 17, the *Feralia* on 21, the *Terminalia* on 23, the *Regifugium* on 24, and the *Equirria* on 27.

September and November were the only months not to have any feast days, with the exception of the Ides. Some feast days—such as the mysterious *Agonalia* on 21 May, 11 December and 9 January—return to fall on different days during the year; others, such as the *Lemuria* on 9, 11 and 13 May, fall on three consecutive odd-numbered days. Incidentally, only the *Equirria* on 14 March and the *Regifugium* on 24 February fall on even-numbered days.

Indeed, even numbers seem to be a distinguishing characteristic of the two calendars: the Romulean calendar has 6 months of 30 days, and lasts for a total of 304 days, while in the Numan calendar February alone has an even number of days, and the year lasts a total of 355 days.[2]

[2] Macrobius, *Saturnalia*, 1.13.5, attributes to both Numa and Pythagoras a predilection for odd numbers.

Virgil, *Bucolica*, 8.75, recalls: "*numero Deus impare gaudet*, God is pleased by odd numbers". See also Joannes Lydus, *de mensibus*, 3.44.

Chapter 3
The Names of Etruscan and Roman Months

An opening assertion: mankind's first ever calendar derived from the Moon and observation of its motion: the Moon making an entire revolution around the Earth inspired the idea of a "lunation", a span of time longer than a single day but relatively short nonetheless. The Indo-European name for "moon"—which has been reconstructed as **menes-* or **me(n)s*—provides the root of our word for "month", in the sense of a lunar month or a lunation. The noun "month/moon" is part of the common heritage shared by Indo-European peoples, which implies that the very idea of a lunation dates back to *before* the separation of these peoples from their—presumed—common homeland.

And yet the Moon's monthly revolution around the Earth—its "lunation"—has no direct, measurable relationship with the Sun's apparent revolution around the Earth. And it is this astronomical cycle which, over the course of time, was set to become more significant to human life.

The fact is that in the first, exclusively lunar calendar, as each solar year elapsed the "lunations" fell at different times of year in terms of solar radiation, meteorological conditions, and the advancement of animal and plant life. The second calendar—the solar, or rather lunar/solar calendar—evolved to cater to the changing needs of man, notably, the transition from hunter gatherer to farmer and agricultural civilization. In practice, the same name, "month", was used to denote a period of time of a slightly different length: no longer the period of natural lunation, which lasts around 29.53 days, but rather a "virtual" lunation, which, at least in Rome, could last either 28, 29, 30 or even 31 days.

Twelve such virtual months—which do not coincide with the lunations—made up a civil year that was almost the same length as the lunar year, but failed to coincide with the solar year at all. For instance, the Roman calendar traditionally attributed to Numa Pompilius consisted of 12 virtual "months" spanning a total of 355 whole days. This is 0.6328 days longer than 12 lunations, but 10.2422 days shorter than the solar year.

L. Magini, *Stars, Myths and Rituals in Etruscan Rome*, Space and Society,
DOI 10.1007/978-3-319-07266-1_3

Intercalation was inevitable to try and reconcile the civil lunar year with the solar year, in order to make sure that the feast days associated with agricultural labour always fell on the same days each civil year, and more or less remained within the same period during the solar year. As a result of this, each "month" in the intercalated lunar/solar year was no longer tied down to the lunation, but rather fell during a given period in the solar year, with its particular solar radiation, meteorological conditions, and advancement of animal and plant life.

Introduction of an intercalated lunar/solar year made it necessary to give specific names to each of the 12 months, rather than using the generic and rather anonymous word "month" for every lunation. Twelve names were allocated to the 12 months, which are the direct predecessors of our modern-day January, February, March, etc. These names served to bind each month to its respective—but still approximate—period during the solar year.

The creation of this type of year and its associated names for months is a relatively recent event that took place *after* the Indo-European diaspora, and the slow, tortuous transition from hunter gatherers to animal and crop farmers. This explains why modern-day names for months in the various Indo-European languages *do not* derive from a common Indo-European root. As Carl Darling Buck explains in *A Dictionary of Selected Synonyms in the Principal Indo-European Languages*:

> Previous to the widespread, though still incomplete, European adoption of the Latin names, there was the utmost diversity. In ancient Greece alone there were dozens of different local calendars (cf. Pauly-Wissowa 10.175 ff.). There was no agreement between the old Germanic names (even the Old High German lists vary somewhat), nor between Indic and Iranian, not even between the Avestan and Old Persian names. Neither the modern Lithuanian literary names, a recent coinage, nor the diverse older and dialectic forms agree with the older, now obsolete, Lettic names. Among the native Slavic names, which are still current except in Russian, there are a few cases of agreement between Bohemian and Polish, and a few others of common words applied to different months (e.g. *listopad*, lit. 'leaf-falling', in Serbo-Croatian 'October', Bohemian and Polish 'November'; Bohemian *kveten* 'May' = Polish *kwiecien* 'April', originally 'flower-month'), but, on the whole, great diversity. In general, the month-names are based upon religious festivals (so most of the Greek and some of the Celtic and Germanic) or upon some characteristic feature of the weather, vegetation, harvest, etc. The enumeration and discussion of these so diverse month-names (even if one chose for the Greek only the Attic or for Old High German only those prescribed by Charlemagne) would require so much space that it seems best here to consider only the Latin names and their spread.[1]

This long and detailed introduction serves to highlight the significance of the fact that four out of eight Etruscan month names correspond to their Old Persian corollaries.

Medieval Latin manuscripts have preserved eight of the names used by the Etruscans, running from March to October. These were handed down through the various codices of the *Liber Glossarum*, which was written prior to the 8th century

[1] Buck 1988, p. 1010.

Table 3.1 Names of months in Old Persian and Etruscan

Old Persian months	Etruscan months (from the glossarums)	Etruscan months (certified or reconstructed)	Roman months	Modern months
adukanis	**Velcitanus**	***velχit(a)na**	Martius	March
θura-**vahara**	Cabreas	*capr-	Aprilis	April
θaig(a)rcis	Ampiles	*anpile	Maius	May
garmapada-	**Aclus**	**Acale**	Iunius	June
***θur(a)na**-bahsis	**Traneus**	**Turane**	Quintilis	July
***garma-**bahsis	**(H)ermius**	***hermi**	Sextilis	August
bagayadis	Celius	Celi	September	September
***varkazana**	Xosfer	*cezpre	October	October

This comparison is restricted to the names of the first 8 months, as these are the only documented names in Etruscan. The names for months in Old Persian are from Brandenstein-Mayrhofer 1964, p. 9; the ending *-bahsis* in the months of July and August, which are reconstructed from their Elamite form, is thought to stand for "month". As noted earlier, Etruscan month names from various glossarums are in Mountdorf 1923, p. 108; certified or reconstructed Etruscan month names are in Caffarello 1975, p. 111. Reconstructed forms are indicated by an asterisk. Phonetically corresponding forms are in bold

CE, and are confirmed in Papia's *Elementarium* from the 11th century CE. In three out of eight cases, surviving documentary evidence is sufficient to make a satisfactory comparison between late medieval information and the corresponding Etruscan original from at least a thousand years earlier, during the centuries immediately prior to the birth of Christ. Furthermore, "from the omission of the months November-Februarius we may conjecture that the list began with March and was already defective when the *Liber Glossarum* was compiled."[2]

The names of the Old Persian months have a rather complex history, which we do not have the scope to cover here in detail. For our purposes, it is sufficient to note that some of the names have been preserved solely through the Elamite language, and have been reconstructed in their original Iranian form.[3]

Table 3.1 presents:

- The names of the Old Persian months;
- The names of Etruscan months handed down through glossarums;
- The names of certified or reconstructed Etruscan months—the latter indicated by an asterisk;

[2] Mountdorf 1923, pp. 108–109, provides and comments on the names of the Etruscan months that have survived through the various glossarums. The names of certified or reconstructed Etruscan months are from Caffarello 1975, p. 111.

[3] See Brandenstein-Mayrhofer 1964, p. 9, and the prior bibliography.

- The names of Roman months;
- The corresponding names of months commonly in use in English.

Comparing the Old Persian and Etruscan forms, four phonetical correspondences are immediately evident:

1. The names for March in Etruscan and October in Old Persian;
2. The second part of the Old Persian name for April and the Etruscan name for June;
3. The names for July;
4. The names for August.

It is worthwhile looking a little more closely at these points of correspondence.

1. *Varkazana*, the Iranian name for "October", is certified in line 88, column 3 of the huge King Darius (550-486 BCE) inscription at Behistan (modern Bisutun). This modified form of the Elamite name for the eighth month of the year, *marqasanas* (*arahsamna* in Akkadian), makes it possible to interpret the name in Iranian as a composite of *varka-* "wolf" and *zana-* "man, human being", to produce the meaning "(month) of the Wolf man".[4] It should also be noted that in Rome, October and March were both "martial" months, marking the beginning and end of the warring season; it is hardly a revelation that Mars is more closely associated with the wolf than with any other animal in creation.
2. The Iranian name for "April"—*θuravahara*—is a composite form. The second element, *vahara-*, is the Iranian form of the Indo-European word for "spring", which becomes *εαρ* in Greek and *ver* in Latin. Now let's turn to Hesychius's lexicon: "boys (are called) **acaletur* by the Etruscans."[5] Looking at a comparison of this word with the Etruscan name for "June", we find a similar explanation for the name *Iunius* in Latin as the month of the *iuniores*, "the young(est)"; whether or not this explanation is true or false is moot. Meteorological conditions may potentially be adduced as the reason why the name for Iranian April became Etruscan June.
3. The first part of the Old Persian name for "July", **θur(a)na*, is probably connected with the Sanskrit *surah* and Avestan *sura-*, meaning "strong". This latter word is one of the names for the Iranian Venus/Aphrodite: *Aredvi*, *Sura*, *Anahita*, "Moist, Strong, Immaculate"; in Etruscan, *Turan* is the name of the goddess of love, the Etruscan Venus/Aphrodite. We should also recall that the Roman word for "July", *Quintilis*, mutated over time to become *Iulius*, from the name of Venus's son's son.
4. The first part of the Old Persian noun for "August", *garma-*, in Iranian means "hot", and corresponds to the Sanskrit *gharmah* and the Latin *formus*. These three nouns—in Iranian, Sanskrit and Latin—have the same semantic value, and all three come from the well-represented Indo-European form **g^{w}hermo-*.

---

4 Kent 1953, p. 206; Brandenstein-Mayrhofer 1964, pp. 151–152.

5 TLE 1968, n. 802: *αγαλητορα παιδα. Τυρρηνοι.*

It is also worth bearing in mind that:

- *Sextilis*/August is in the heart of the "hot season", the "summer", for which the Latin is *aestus,-us*, "burning heat";
- *garma-* in Iranian, is the opposite of *aota-*, "cold";
- *aestas,-atis*, "summer, the hot (season)", in Latin is followed and contrasted by *autumnum* (sc. *tempus*), "autumn, the cold season";
- *autumnus* " is probably Etruscan in origin" (Ernout-Meillet, s.v.).

At least a further two more significant elements may be added to these four phonetic and semantic correlations regarding the names of the months.

The first is that the name of the seventh month in the Old Persian year, *bag-ayadis*, is composed of *baga-*, "God", and *yad-*, "worship", which allocates the ancient Iranian month of September/October the value of "(month) of worshipping God". The God in question is, in all likelihood, Mithra, as in Pahlavi and Persian the month is called *Mihr*, "(the month of) Mithra."[6] In Rome, 13 September was the most important feast day of the year: *Epulum Iovis*, "Jupiter's Banquet", was the day for worshipping Jupiter Optimus Maximus, the Etruscan/Roman counterpart to Mithra in Iran.

The second significant element, both in Rome and in Iran, is that 21 April is not just any old day: it is "mid-spring day". In consequence, there is a "solar" New Year's Day, both in Rome and Iran, on 9 March (see Chap. 7).

If all of these comparisons are correct, we may start drawing some conclusions. The most important, at this stage, is that if there is an unbroken line linking the names of the months—the Roman names were copied from Etruscan names, the Etruscan names copied from Old Persian, and the Old Persian names carried over from the Mesopotamian names—isn't this a clue to the distant Eastern origins of the Roman calendar? Wouldn't it be about time to start acknowledging Etruscan culture—at least, Etruscan astronomical knowledge—as an intermediary between the Orient and Rome?[7]

The explanations we shall see later on regarding the solar New Year, the cult of Venus, the cycle of Saros and, above all, the Roman festivities of *Regifugium* and *Vestalia*, all support this interpretation of available sources.

---

[6] Kent 1953, p. 199.

[7] On other aspects of the relationship between the Orient and Etruscan civilization, see Magini 2007, Magini 2008 and Magini 2011.

Chapter 4
The Numan Cycle

Every other year, the 355 day Numan year was intercalated by 22 or 23 days: odd-numbered years of 355 days alternated with even-numbered years of 377 or 378 days. In the third century, Censorinus writes:

> Afterwards, Numa […] created 12 months and 355 days, though he was aware that the Moon completes 12 lunations in 354 days […] When it was decided to add a 22- or 23-day intercalary month in alternate years to make the civil year the same length as the natural year, a preference was expressed for February, between the *Terminalia* and *Regifugium*; and so it came to pass for many years before anybody noticed that civil years were becoming notably longer than natural years[1].

It is perfectly straightforward to calculate how much longer civil years were than natural years: using the aforementioned intercalation, 8 years last 2,930 days,[2] eight more than the 2,922 days[3] that make up 8 solar years. This blunder defeated the purpose of the intercalation; namely, to reconcile the lunar and solar years, and the solar year with the civil year.

---

[1] Censorinus, *de die natali*, 20.4-6: *Postea […] a Numa […] XII facti sunt menses et dies CCCLV, quamvis luna XII suis mensibus CCCLIIII dies videbatur explere […] Denique cum intercalarium mensem viginti duum vel viginti trium dierum alternis annis addi placuisset, ut civilis annus ad naturalem exaequaretur, in mense potissimum Februario inter Terminalia et Regifugium intercalatum est, idque diu factum prius quam sentiretur annos civiles aliquanto naturalibus esse maiores.*

The *Terminalia* fall on 23 February, and *Regifugium* on 24 February.

Further comments on the Numan calendar and its intercalation may be found in Varro, *de lingua latina*, 6.13 and Plutarch, *Life of Numa*, 18.

[2] $2{,}930 = (8 \times 355) + 22 + 23 + 22 + 23$.

[3] $2{,}922 = 8 \times 365.25$.

L. Magini, *Stars, Myths and Rituals in Etruscan Rome*, Space and Society,
DOI 10.1007/978-3-319-07266-1_4

Writing in the fifth century, Macrobius explains:

> When this error was discovered too, the following correction was adopted: every three eight-year cycles were allotted not 90 but 66 intercalated days, to offset the 24 extra days that accreted over 24 years.[4]

This late and isolated statement[5] has been roundly ignored up to the present day, and yet it is crucial, as it clarifies that Numa's "error" was to be corrected over a 24 year cycle, sub-divided into three sub-cycles: the first two 2,930 days long (8 Numan years of 355 days plus 90 intercalary days, distributed as stated above), and a third 2,906 days long (consisting of 8 years lasting 355 days plus 66 intercalary days). For the sake of simplicity, it is easiest to refer to this as the *Numan cycle*, though Macrobius fails to attribute it explicitly to the second king of Rome.

Macrobius further fails to specify how the 66 days intercalated into the final 8 years of the cycle are divided up; nor does he mention the specific years when they were inserted. To fill these gaps, we must make *two hypotheses*:

- as in the first two sub-cycles, the 66 intercalatory days were divided into three groups of 22 days each;
- the 22 days were inserted in the 3rd, 6th and 8th year of the final sub-cycle, as practiced in cultures such as Babylon and later Athens,[6] both of which adopted 8 year cycles.[7] Table 4.1 is a reconstruction of the *Numan cycle* following these two hypotheses.

The Numan cycle lasts 8,766 days: 13 years of 355 days plus 7 years of 377 days and 4 years of 378 days, that is to say 24 lunar years of 355 days plus 246 intercalary days, at *an average of 10.25 (= 246 ÷ 24) intercalary days per year*. The cycle corresponds to three times the Athenian octaeteride of 2,922 days.

---

[4] Macrobius, *Saturnalia*, 1.13.13: *Hoc quoque errore iam cognito haec species emendationis inducta est. Tertio quoque octennio ita intercalandos dispensabant dies, ut non nonaginta sed sexaginta sex intercalarent, compensatis viginti et quattuor diebus pro illis qui per totidem annos supra [...] numerum creverant.*

[5] Macrobius wrote in the fifth century CE; however, his sources were sound: Svetonius's *de anno Romanorum* and Cornelius Labeo's *Fasti*. See Marinone 1987, p. 45.

[6] According to the methods set forth in the *Ars Eudoxi* (see Geminus's *Introduction to the Phenomena*, 8.33; no. 1, p. 53 in the Les Belles Lettres edition, edited by G. Aujac): Geminus posits that the intercalation took place in the 3rd, 5th and 8th years. In actual fact, given that by the 16th year the cycle is more than 16 days ahead of the solar year (16 × 365.2422 = 5,843.8752; 5,860 − 5,843.8752 = 16.1248), it would be more reasonable for the intercalated years in the third subcycle to be the 4th, 6th and 8th years (see Chap. 15, no. 1).
Another intercalation model would be needed to reconcile the 24 year Numan cycle cited by Macrobius with the 19 year Metonic cycle noted by Livy in 1.19.6. Neither of these options affects the following calculations, given that the astronomical cycles at issue last less than 19 years; this is an issue that requires further research.

[7] The eight-year cycle—known in Greek as an octaeteride—equates 8 solar years (in modern figures: 2,921.9376 = 365.2422 × 8) to 99 lunations (2,923.5294 = 29.5306 × 99).

Table 4.1 The Numan cycle

Sub-cycles	Cycle years	Days in the year	Intercalary days	Total days in the year	Total days in the Numan cycle
First sub-cycle	1	355		355	355
	2	355+	22=	377	732
	3	355		355	1,087
	4	355+	23=	378	1,465
	5	355		355	1,820
	6	355+	22=	377	2,197
	7	355		355	2,552
	8	355+	23=	378	2,930
Total first sub-cycle		2,840+	90=	2,930	
Second sub-cycle	9	355		355	3,285
	10	355+	22=	377	3,662
	11	355		355	4,017
	12	355+	23=	378	4,395
	13	355		355	4,750
	14	355+	22=	377	5,127
	15	355		355	5,482
	16	355+	23=	378	5,860
Total second sub-cycle		2,840+	90=	2,930	
Third sub-cycle	17	355		355	6,215
	18	355		355	6,570
	19	355+	22=	377	6,947
	20	355		355	7,302
	21	355		355	7,657
	22	355+	22=	377	8,034
	23	355		355	8,389
	24	355+	22=	377	8,766
Total third sub-cycle		2,840+	66=	2,906	
Total Numan cycle		8,520+	246=	8,766	

Two points (which we will be examining in greater depth later) are worth noting:

- The Numan cycle is asymmetric: the first 12 years last 24 days longer than the last 12 years, but *the first 11 years*, lasting 4,017 days, *are close to eleven solar years and to 136 lunations*[8] (see Chap. 21);
- Up to year seventeen (inclusive), with the standard intercalation, *10 years*, corresponding to 3,662 or 3,663 days, *are close to 124 lunations*[9] (see Chap. 15).

[8] 4017.6642 days = 11 × 365.2422; 4016.16161 days = 136 × 29.5306.
As we shall see (in Chap. 21, especially no. 2), the reason why the Numan cycle is not made up of three equal portions numbering 2,922 days each, obtained by intercalating alternately 20 and 21 days during even-numbered years, is precisely this: *the first 11 years of the cycle must maintain the same number of days as 11 solar years.*

[9] 3,661.7944 days = 124 × 29.5306.

Chapter 5
The Numan Year and Cycle, and the Motion of the Moon

The intercalations in the Numan cycle disrupt any relationship between the Moon's phases and the day of the month. The Calends were *not* set by the Pontifex Minor's—i.e. younger—first sighting of the crescent Moon; the Nones *do not* indicate days when the Moon reaches the first quarter; and the Ides *do not* correspond to the full Moon.[1]

The months of the Numan year also fail to mirror lunations:

- If the first crescent Moon (usually 1.5 days[2] after the new Moon) is spotted at the beginning of a 31 day month, at the next Calends the Moon is 2.97 days old, and already more than 36° from the Sun;
- If the month is 29 days long, at the next Calends the Moon is 0.97 days old, and just 11.8° from the Sun (and far too close to be observed);
- If the month is 28 days long, at the next Calends the Moon is in conjunction with the Sun. This is a particularly propitious astral configuration for relations between the micro- and macrocosms, which may explain why February, the only month that is 28 days long, is dedicated to the *Manes*[3]—i.e. to those mortals who (temporarily) reside in the macrocosm.

---

[1] Contrary to Macrobius's assertion in *Sat.* 1.15.9 and 15.

[2] This value, corresponding to a distance of around 18° (= 1.5 × 12.1908°) from the Sun, is the one adopted in Babylon (Schiaparelli 1997–1998, Vol. III, pp. 121–233; the figure—on p. 225—is taken from Epping-Strassmeyer's *Astronomisches aus Babylon*, pp. 18–24, and derives from more than 30 observations during all seasons of the year covered therein). In reality, the interval between conjunction and the "apparent new Moon", as Schiaparelli refers to the first visible crescent, is one of the most variable, predominantly owing to the irregularity of lunar motion, the season, and the latitude from which the observation is being undertaken. Nevertheless, the indicated value matches astrological tradition sufficiently closely: "the moon 'rises' when it is 15° from the sun", writes Rhetorius, *de planetarium natura ac vi* (see CCAG VII, pages 213–224; cited in Bezza 1995, vol. 1, pp. 72–86).

[3] On February, the month of the *Manes*, see Macrobius, *Saturnalia*, 1.13.3 and Joannes Lydus, *de mensibus*, 3.10.

The figures in the text are the result of the following calculations: 2.97 days = 32.50 − 29.53; 36° = 2.97 × 12.1908°; 0.97 days = 30.50 − 29.53; 11.8° = 0.97 × 12.1908°; 29.53 days ~ 1.5 + 28 days.

L. Magini, *Stars, Myths and Rituals in Etruscan Rome*, Space and Society,
DOI 10.1007/978-3-319-07266-1_5

However, an understanding of the Moon's synodical motion may be inferred from the 16 day interval between the (theoretical) full Moon, the Ides, and the first crescent (once again, theoretical) of the following Calends, which is the same for all months save February.[4] After 16 days, the Moon is approximately 195° from the Sun;[5] if, on the first of these 16 days, the practically full Moon has just emerged from opposition with the Sun, 16 days later it is just 15° beyond conjunction with the Sun, and therefore about to reappear as a crescent.

According to the authors of antiquity, the reason why the lengths of months and of the year were rounded up was a preference for odd numbers over even numbers.[6] In actual fact, the twelve synodic months total 354.3672 days, while thirteen sidereal months equal 355.1808: the 355 days of the Numan year are closer in length to thirteen sidereal months than twelve synodic months. If the year lasted 354 days, at the following new year the Moon would be 0.367 days younger than it had been on the preceding new year. If we establish the start of the year with the appearance of the first crescent at the Calends of March, by the following new year's day the Moon would be 1.133 days old: just 13.8° from the Sun, the earliest crescent would still be invisible and the year would not have been deemed to have commenced.[7]

And yet the first eleven months of the Numan year total 327 days: 1 February is the 328th day after 1 March. It follows that on 1 February the Moon will have completed almost exactly twelve sidereal revolutions; at the same time of night, it is in the same position as it was with respect to the fixed stars on the preceding 1 March.[8] This is a clear demonstration of the understanding of the Moon's sidereal motion.

Intercalated Numan years did not respect lunations either:

- In 355 days, the Moon travels 4,677.62° and is 2.38° from completing its 13th sidereal revolution; it is 0.633 days old in the 13th synodic revolution, and around 7.7° from the Sun;[9]—In 377 days, the Moon travels 4,967.5° and is

---

[4] Macrobius tells us in *Saturnalia*, 1.15.7: "It was believed correct that, every month, the Ides fell on the 9th day after the Nones, and it was thus established that 16 days should elapse between the Ides and the following Calends, *Omnibus tamen mensibus ex die nonarum idus nono die repraesentari placuit, et inter idus ac sequentes kalendas constitutum est sedecim dies esse numerandos.*"

[5] 195.0528° = 16 × 12.1908°.

[6] On odd and even numbers, see, among others, Censorinus, *de die natali*, 20.4.

[7] These figures were calculated as follows: 354.367 days = 12 × 29.5306; absolute error +0.633; percentage error 0.179 %; 355.181 days = 13 × 27.3216; a.e. −0.181; p.e. 0.051 %. 0.367 days = 354.367 − 354; 1,133 days = 1.5 − 0.367; 13.8° = 1.133 × 12.1908°.

[8] These figures were calculated as follows: 12 × 27.3216 = 327.8592 days.
For Mayan knowledge of an 82-day cycle (= ¼ of 328 days), see Aveni 2000, p. 47.

[9] The figures were calculated as follows: 4,677.62° = 355 days × 13.1764°; 4,680° = 13 × 360°; 2.38° = 4,680° − 4,677.62°; 0.633 days = 355 − 354.367; ca. 7.7° = 0.633 × 12.1908°.
From hereon in, only the results of similar calculations will be presented.

72.5° from completing its 14th sidereal revolution; it is 22.633 days old in the 13th synodic revolution, and around 276° from the Sun, just past negative quadrature;[10]

- In 378 days, the Moon travels 4,980.68° and is 59.3° from completing its 14th sidereal revolution; it is 23.633 days old in the 13th synodic revolution, and around 288° from the Sun, in the negative quintile.[11]

From this data, it is possible to approximate the movement of the Moon with respect to the fixed stars and the Sun from one year to the next, depending on how long the preceding year lasted.

At the end of the entire 8,766-day-long Numan cycle, the Moon has travelled just 304° of the 321st sidereal revolution, 305° of the 297th synodic revolution, and is around 25 days old. It completes the 321st sidereal revolution and the 297th synodic revolution almost simultaneously: the former at 8,770.2336 days and the second at 8,770.5882 days[12] after the beginning of the cycle.

A revised form of intercalation between one Numan cycle and the next, or between multiple Numan cycles, was required for the Moon's motion to be restored to harmony with respect to the Sun's movement.

---

[10] $-84° = 276° - 360°$.

[11] $-72° = 288° - 360°$.

[12] $8{,}770.2336 = 321 \times 27.3216$; $8{,}770.5882 = 297 \times 29.5306$.

Chapter 6
The Numan Cycle and the Movements of the Sun, the Moon and Venus

To some degree of precision, the 8,766 day Numan cycle may be reconciled with the movements of the Sun, the Moon and Venus (see Table 6.1).

Table 6.1 shows that the Numan cycle in particular respects the movement of the Sun. It is sufficiently precise for the civil year to gain less than one day on the solar year over the course of five cycles, equivalent to 120 years.[1] In actual fact, the margin of error with regard to the Sun's motion—ca. 4 h 33′ 36″ every 24 years—is 2.5 × less than that of the Metonic cycle,[2] which was introduced in Athens in 432 BCE; the margin of error with regard to the Moon's motion is around 12 times greater. The margin of error with regard to the motion of Venus is a little higher without being excessive.

In conclusion, the individual years of the Numan cycle—which are either 355, 377 or 378 days long—do not correspond to solar years; the individual months in the Numan year—which are either 31, 29 or 28 days long—do not correspond to the synodic months; however, as a whole, the entire 8,766 day Numan cycle corresponds very well to the motion of the Sun, if less well with the motion of the Moon. As for Venus, it is relatively easy to associate its location with respect to the Sun (see Chap. 9) with the festivities of the Numan year over an entire Numan cycle—that is to say, during 15 synodic periods of the planet—through an approximation comparable to that of the Numan years and months with respect to solar years and lunar months.

Schiaparelli explains in his commentary on data discovered on Mesopotamian tablets referring to the motion of Venus, in which he distinguishes between *Observations* (recordings of phenomena observed) and *Ephemeris* (the results of "purely arithmetic labour"):

---

[1] $8{,}766 \times 5 = 43{,}830$ days; $365.2422 \times 120 = 43{,}829.064$; $43{,}830 - 43{,}829.064 = 0.936$ days.

[2] The Metonic cycle equates 19 solar years with 235 lunations. In actual figures, $19 \times 365.2422 = 6{,}939.6018$ days, $235 \times 29.5306 = 6{,}939.691$.

L. Magini, *Stars, Myths and Rituals in Etruscan Rome*, Space and Society,
DOI 10.1007/978-3-319-07266-1_6

Table 6.1 The Numan cycle and the movements of the Sun, the Moon and Venus

Celestial body	No. of cycles × period = days	Absolute error	Percentage error (%)
Sun (tropical year)	24 × 365.2422 = 8,765.81 days	+ 0.19 days	0.00217
Moon (synodic month)	297 × 29.5306 = 8,770.59 days	− 4.59 days	0.0523
Moon (sidereal month)	321 × 27.3216 = 8,770.23 days	− 4.23 days	0.0482
Moon (draconitic month)	322 × 27.2122 = 8,762.33 days	+ 3.67 days	0.0419
Moon (anomalistic month)	318 × 27.5546 = 8,762.36 days	+ 3.64 days	0.0415
Venus (synodic period)	15 × 583.92 = 8,758.8 days	+ 7.2 days	0.0822
Venus (sidereal period)	39 × 224.63 = 8,760.57 days	+ 5.43 days	0.062

> In all likelihood, I would say that the *Ephemeris* were based on more imperfect elements and made prior to the *Observations*: this would have given observers the background to make observations themselves, by means of an approximate prediction for the period in which they could expect a given phenomenon to occur. The same procedure continues to be used on a regular basis by modern observers. Without spadework of this sort, it would be impossible, or at least extremely laborious and difficult, to undertake the precise observation of an eclipse or the return of certain comets, not to mention the passage of a heavenly body over the Meridian.[3]

To recap: the Roman festivities during the age of its earliest kings—the so-called *Feriale antiquissimum* (see Chap. 2)—were the *Ephemeris* of the nameless inventor of the Numan year and cycle, as the following calculations bear out.

[3] Schiaparelli 1997–1998, Vol. III, p. 155.

Chapter 7
The Solar New Year on March 8 or 9

The intercalary days were inserted after the feast of *Terminalia* on 23 February. The feast was thus named

> because this day is set as the last day of the year; for the twelfth month was February, and when the extra month is inserted the last five days are taken off the twelfth month.[1]

The only logical conclusion we may infer from these three facts—(1) that the intercalation helped to offset the difference between the lunar and solar years; (2) the solar year is 365 days long; and (3) 22 or 23 intercalated days are inserted after the "last day" of the lunar year—is that *the solar year ending with the 22nd or 23rd intercalated day begins on the preceding 8 or 9 of March.*[2]

A number of things change if the solar year begins on 8 or 9 March, and the new Moon or first crescent coincides with New Year's day. These modifications clear up at least three issues:

1. 45 days later, on 21 April, is the *Parilia* and Foundation of Rome festival, when "springtime is half run"[3];

---

[1] Varro, *de lingua latina*, 6.13: [...] *quod is dies extremus constitutus: duodecim enim mensis fuit Februarius et cum intercalatur inferiores quinque dies duodecimo demuntur mense.*

The expression *cum intercalatur* should be translated as "every time that the intercalation is undertaken", rather than "when the extra month is inserted". We shall see later (see Chap. 16) how, why and in what way "the final 5 days are taken off".

[2] In my book *The Feasts of Venus* ("Le feste di Venere") I merely considered the 8th of the month. I have since realized that this line of reasoning applies to both days, considering that for the Romans the day started at midnight, while the Etruscans counted their days from midday (see Servius, *Ad Aen.* 5.738 and 6.535).

9 March is an *ancilia moventur* day (see Chap. 22), and shares the same nundinal letter (A) as day 1.

[3] Columella, *de re rustica*, 11.2.36: *ver bipertitur.*

21 April is the feast of "mid-spring" both in Rome and in Ancient Iran, where the year began on 8 March and the months were 30 days long. See Bartholomae 1961, s.v. *maδyoi.zaremaya-* "mid-spring", *yairya-* "anno" and *maiδyoi.sam-* "mid-summer".

L. Magini, *Stars, Myths and Rituals in Etruscan Rome*, Space and Society,
DOI 10.1007/978-3-319-07266-1_7

2. On the Nones of July, four lunations after the solar new year, comes another new Moon, at which "an eclipse of the Sun with the *Nonae Capratinae*, a day still reserved for solemn celebration"[4] may be observed; Plutarch mentions this when recalling the celestial event that occurred on Romulus' death;
3. The new Moon of the Nones of July also makes it clear why we hear assertions such as "Some (say) that the Nones (is so called) for the new Moon, because they mark the start of the lunation", or "The Nones are thusly called […] because, just as the Calends of January are known as the 'new year' of the new Sun, they are known as the 'new month' of the new Moon."[5]

If the calendar year was not intercalated, the solar year beginning on 8 March would end on 17 March the following year, during the feast of the *Liberalia*. This was the moment when the God *Liber*, i.e. the Sun,

> as the days grow longer, at the spring equinox, increases his energies like an adolescent, and is portrayed in the guise of a young man.[6]

On that same day, the cycle of adolescence drew to a close for Roman males, marking the beginning of *iuventus*, or "youth". With *tirocinium* or "training" now over, they donned the manly toga, were eligible for the army and, most importantly, were presented to the population—on the feast day of 17 March—as *Liberi*, or "free men", who had come of age.

If we do the sums, it becomes apparent that the new Moon falls on around 8 or 9 March, while the first crescent appears around 9 or 10 March, only in years 6, 15 and 16 of a Numan cycle that commenced with a new Moon or, better, with a total eclipse of the Sun on 1 March in year one (see Chap. 1). The last of these years matches Aulus Gellius's statement well:

> Servius Tullius, the king […] considered male sons up to the age of 16 to be 'boys'; from 16 onwards, he believed them old enough to serve the State, enlisted them as soldiers […][7]

---

[4] Plutarch, *de fortuna romanorum*, 320C: *λεγουσι γαρ εκλειποντος του ηλιου ηφανισθαι, νωναις καπρατιναις, ην αρχι νυν ημεραν επιφανως εορταζουσιν.*

[5] P. Fest., p. 176L: *nonas quidam a nova luna, quod in eas concurreret principium lunae.*; Varro, *de lingua latina*, 6.28: *nonae appellatae […] quod, ut novus annus Kalendae Ianuariae ab novo sole appellatae, novus mensis ab nova luna Nonae.*

[6] Macrobius, *Saturnalia*, 1.18.10: *[…] procedentibus augmentis aequinoctio vernali similiter atque adulescentis adipiscitur vires figuraque iuvenis ornatur.*

[7] Aulus Gellius, *Noctes Atticae*, 10.28: *[…] Servium Tullium regem […] pueros esse existimasse, qui minores essent annis septem decem, atque inde ab anno septimo decimo, quo idoneos iam esse reipublicae arbitraretur, milites scripsisse […]*

The roman practice of counting inclusively increases the result by one compared with our favoured method of counting: as a result of this, the expression *minores […] annis septem decem* should be translated as "up to the age of sixteen".

Chapter 8
The Movements of the Moon and Venus and the Language of Myth

The Goddess *Fortuna*—who was imported to Rome by her protégé Servius Tullius—has lunar characteristics (Fig. 8.1): she is as nocturnal, transitory and changeable as "the night jewel".[1] It was generic "great good fortune" that the "conception and generation of Romulus" engendered "a solar eclipse, with the Sun in exact conjunction with the Moon for the length of time that the God Mars remained in conjunction with the mortal Silvia" (see Chap. 1). It was also the Goddess Fortuna in person who, at night, awakened the Capitoline geese; it was this selfsame Goddess who visited the slave-turned-king Servius Tullius, "dropping down into his room through a small window", and who "was wont to enter his house at night-time by a small window, hence the gate bears the name of *Fenestella* ("the Little Window")."[2] And that's not the half of it.

The incredibly complex motion of the Moon (see Chaps. 17, 18 and 19) goes through one particularly complicated moment when, as it wanes, the Moon inexorably closes in on the Sun, goes into conjunction with the new Moon, and re-emerges as the first crescent. In the language of myth, this rapid succession of the Moon's changing aspects is embodied by three divine figures: *Fortuna* and her two handmaidens, *Necessitas*, who precedes her, and *Spes*, the last of the Goddesses.

They are represented in the heavens respectively by the new Moon, the waning Moon and the waxing Moon, and in Ancient Rome replicated the three Greek Graces (Fig. 8.2). *Fortuna*, the central figure in the group, turns her back—the *caecum corpus*—and, more specifically, the rotundities which, in the Eternal city, to this day, continue to be highly esteemed as a talisman. In Rome, *caeca nox* was

---

[1] *Ratri-mani* is one of the many ways of referring to the Moon in Sanskrit.

[2] Plutarch, *de fortuna romamorum*, 320 BCE; see no. 1.2.
The geese awakened by Fortuna in Plutarch, *de fort. Rom.*, 325 BCE.
Fortuna, the window and Servius Tullius in Plutarch, *de fort. Rom.*, 322E: *δια τινος θυριδος καταβαινουσαν εις το δωματιον, ο νυν Φενεστελλαν πυλην καλουσιν*, and in Ovid, *Fasti*, 6.577-8: *nocte domum parva solita est intrare fenestra; / unde Fenestellae nomina porta tenet.*

L. Magini, *Stars, Myths and Rituals in Etruscan Rome*, Space and Society,
DOI 10.1007/978-3-319-07266-1_8

Fig. 8.1 *Fortuna*/Tyche, detail Marble, II-III century CE, *Tomis*/Constanta–Constanta Museum of National History. Photo by Cristian Chirita–Wikimedia Commons

Fig. 8.2 The three Graces, or *Fortuna* and her handmaidens. In the middle, with her back to us, *Fortuna*, represented in the sky by the New Moon; on each side are her two handmaidens, *Necessitas*, the waning Moon, and *Spes*, the waxing Moon (from *Illustrazione de' Monumenti Scelti Borghesiani* di E. Q. Visconti, Roma 1821)

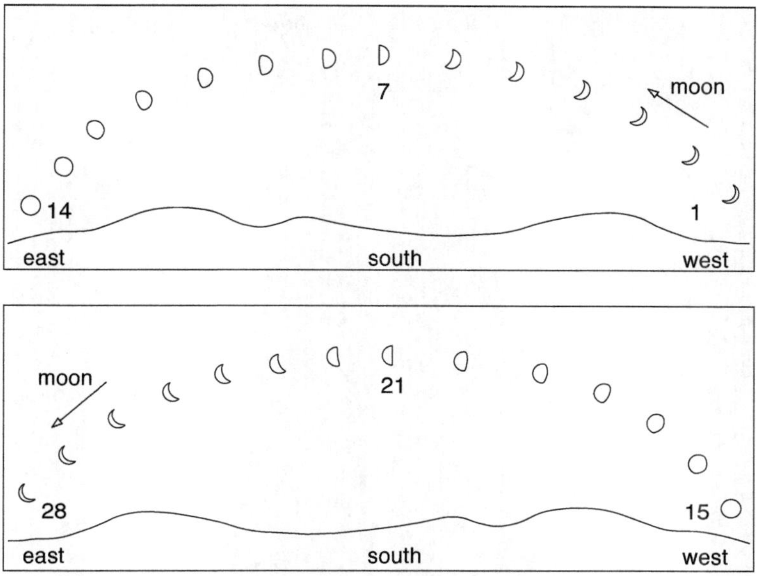

Fig. 8.3 The moon waxing and waning over an entire lunation. Visible to the West after sunset during the first 14 days after lunation, the waxing Moon follows the Sun over the horizon and represents Fortuna *Obsequens*; visible to the East before dawn during the final 14 days of a lunation, the waning Moon faces the part of the sky from which it comes, and represents Fortuna *Respiciens* (from Aveni 2000)

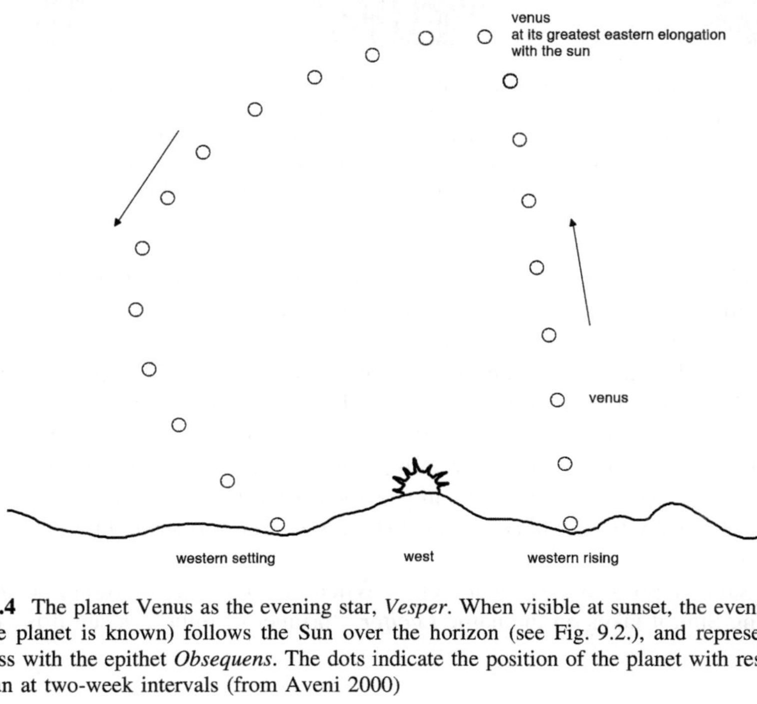

Fig. 8.4 The planet Venus as the evening star, *Vesper*. When visible at sunset, the evening star (as the planet is known) follows the Sun over the horizon (see Fig. 9.2.), and represents the goddess with the epithet *Obsequens*. The dots indicate the position of the planet with respect to the Sun at two-week intervals (from Aveni 2000)

Fig. 8.5 Venus *Callipyge* posing as Fortuna *Respiciens.* With its retrograde motion, the planet retraces the path it has already travelled (see Fig. 8.6), and represents the goddess with the epithet *Respiciens*, "She who must not forget whence she came to become what she currently is". Roman marble from Greek original, 1.52 m in height, from Nero's Domus Aurea in Rome—Farnese Collection, Museo Archeologico Nazionale, Napoli Courtesy Daniele Minopoli Editore, Naples

the "dark night" over which *Fortuna caeca* watched, the "dark Moon", the "Moon that fails to illuminate", or in other words, the new Moon.

The waxing Moon was also known as *Obsequens*—literally "that which goes behind", and figuratively as "obsequious, obedient, propitious"—because only the waxing Moon placidly follows the Sun to dip below the horizon at sunset. The waning Moon acquires *Fortuna*'s other epithet, *Respiciens*, "that which looks backwards": night after night, the waning Moon alone moves from west to east, its "hump facing east" while its face and gaze points westwards, backwards, towards the part of the sky from whence it came (Fig. 8.3). The former and the latter are a mirror image.

The same applies to the planet Venus. When visible in the morning, preceding the rising Sun, it takes on the name *Lucifer*, "bringer of light". When it is visible

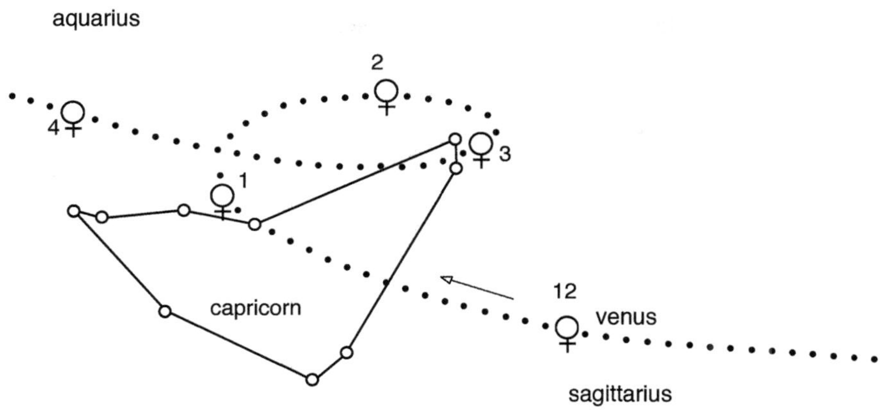

Fig. 8.6 The apparent motion of Venus between December 1957 and April 1958. The planet passes through the same arc in the sky three times: twice in direct motion, once in retrograde motion (from Stumpff 1963)

in the evening, following the setting Sun, it takes on the name *Vesper*, or "evening (star)". In this latter case, the planet behaves like the waxing Moon, as the heavenly representative of the divinity, and is paired with the epithet *Obsequens* (Fig. 8.4).

When, on the contrary, the Goddess looks behind herself—more or less as *Fortuna Respiciens* does—it takes on the name *Kallipyge*,[3] "having shapely buttocks" (Fig. 8.5), and is represented in the sky by the planet which, in its retrograde or clockwise motion (Fig. 8.6), retraces a portion of the path it has just taken.

[3] Torelli 1984, p. 126, recalls a passage in Dio Cassius (42.26.4) in which he describes a statue of Fortuna "which [...] must see and consider everything that is in front of its eyes and behind too, lest she forget her origins, the place from which she began before she became what she is today," noting that "*Fortuna Respiciens* coincides particularly well with the Aphrodite *Kallipyge* type".

Chapter 9
The Motion of Venus

The planet Venus is the brightest body in the sky after the Sun and the Moon. Venus is an internal planet, that is to say, its entire orbit lies closer to the Sun than the Earth's.

Viewed from the Earth, Venus is in one of four configurations specific to the planet (Fig. 9.1):

- in *lower conjunction* with the Sun, when the planet is between the Earth and the Sun, and is not visible;
- at its *greatest western elongation* with the Sun, when the planet is at its greatest angular distance from the Sun to the West, and is visible in the morning, to the East, before the Sun rises;
- in *higher conjunction* with the Sun, when the planet is behind the Sun, and is not visible;
- at its *greatest eastern elongation* with the Sun, when the planet is at its greatest angular distance from the Sun to the East, and is visible in the evening, to the West, after the Sun sets.

Venus completes a full revolution around the Sun—in astronomical terms, its *synodic period*—in roughly 584 days. The planet takes 71 days to move from lower conjunction to its greatest western elongation; it takes 221 days to move into higher conjunction; 221 days later it moves from higher conjunction to its greatest eastern elongation, and then 71 days after that, it reaches lower conjunction: $71 + 221 + 221 + 71 = 584$ days.

The planet's revolution around the Sun may be described in other terms (Fig. 9.2):

- around 5 or so days after lower conjunction, the planet first appears to the East in the early morning sky—known as its *heliacal rising*—and remains visible for around 255 days: preceding the Sun at dawn, like *Lucifer*, it moves away to the West, before drawing in closer once more;

L. Magini, *Stars, Myths and Rituals in Etruscan Rome*, Space and Society,
DOI 10.1007/978-3-319-07266-1_9

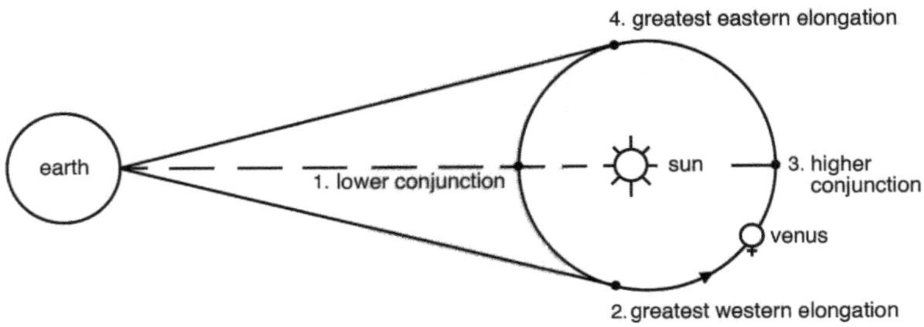

Fig. 9.1 The motion of Venus around the Sun viewed from the Earth. Viewed from the Earth, the planet Venus takes on four specific configurations with respect to the Sun: (*1*) lower conjunction; (*2*) greatest western elongation; (*3*) higher conjunction; and (*4*) greatest eastern elongation

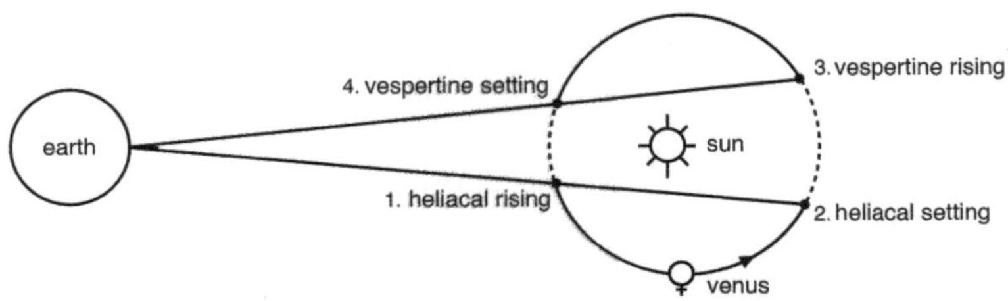

Fig. 9.2 The motion of Venus around the Sun viewed from the Earth. Viewed from the Earth, the planet Venus: (*1*) around 5 days after lower conjunction appears to the East in the morning sky—*heliacal rising*—and remains visible as *Lucifer* for around 255 days; (*2*) disappears—*heliacal setting*—behind the Sun on average for around 62 days; (*3*) reappears to the West in the evening sky—*vespertine or achronic rising*—and remains visible as *Hesperus* for around 257 days; (*4*) disappears once again—*vespertine setting*—around 5 days before lower conjunction, on completion of its synodic period

- it then disappears—its *heliacal setting*—behind the Sun for, on average, 62 days during higher conjunction;
- after this, the planet reappears to the West in the evening sky—at its *vespertine* or *achronic rising*—where it remains visible for around 257 days: following the Sun at sunset, like *Hesperus*, it moves away to the East, before drawing in closer once more (see Fig. 8.4);
- it then disappears once more—its *vespertine setting*—around 5 days before lower conjunction, until it completes its synodic period: $5 + 255 + 62 + 257 + 5 = 584$ days.[1] And so the cycle continues, from one synodic period to the next.

[1] These values are a rounded average of modern values for five consecutive synodic periods of Venus. Mesopotamian tables present very similar values for the analogous positions of Venus (see Swerdlow 1998, pp. 218–219).

At the end of five successive synodic periods, during which, viewed from the Earth, the planet makes a series of five different but equal elegant sweeps as the Morning Star and the Evening Star (Fig. 9.3),[2] Venus returns to its initial position with respect both to the Sun and the fixed stars (Fig. 9.4). This is because the roughly 584 days that make up Venus's synodic period are in a ratio of 8 ÷ 5 with the roughly 365 days of the solar year.[3]

In consequence, at the end of eight solar years, just as Venus is completing its fifth synodic period and is returning for the fifth time to its initial position with respect to the Sun, the Sun is returning for the eighth time to its initial position with respect to the fixed stars, and Venus consequently also returns to its initial position with regard to the fixed stars (see Table 6.1).[4] This is the reason for the *Venus pentagram* (Fig. 9.5), the five-pointed star that indicates the five constellations of the Zodiac and the five moments during the year when, viewed from the Earth, the planet returns to the same configuration.

The history of the pentagram stretches back at least as far as Mesopotamia in 3500 BCE and the earliest ideographic writings under Uruk IV; it reached Anatolia in around 2100 BCE, Minoan Crete in around 1800 BCE, Greece and Etruria in seventh-century BCE, Kazakhstan, Southern Italy, Macedonia and Egypt soon after, and so on (Figs. 9.6, 9.7, 9.8, 9.9, 9.10, 9.11, 9.12, 9.13, 9.14, 9.15 and 9.16).[5]

The five-pointed star earned a place alongside the Islamic crescent Moon. It features on countless national flags, of which the US flag is the best known. It serves as the heroic "star" in the Soviet Union, may be found on the star-shaped epaulettes of Italian military uniforms, and is the so-called *Stellone*, "the Big Star", shining at the centre of the banner of the Italian Republic, keeping a watchful eye on the nation's destiny.

As early as in Babylonian times, Venus the Evening Star was contrasted with Venus the Morning Star. The former represented a feminine deity who promoted

---

[2] The difference in these motions arises from the way in which the planet and the Earth move, and the way that the plane of Venus's orbit is inclined with respect to the plane of the Earth's orbit.

[3] $5 \times 584 = 8 \times 365 = 2{,}920$.

[4] In other words, the 584 days of Venus's synodic period correspond to around (584 ÷ 365.25 =) 1.6 years or (1.6 × 12 =) 19.2 months. It follows that on completion of one of Venus's synodic periods, viewed from the Earth, the planet makes a full revolution of all 12 signs of the Zodiac and starts another revolution of (19.2 − 12 =) 7.2 signs, equal to (7.2 × 30° =) 216°.
At the end of the fifth synodic period, after around (584 × 5 =) 2,920 days, the number of signs of the Zodiac it has traversed is now (7.2 × 5 =) 36, and the number of degrees (216° × 5 =) 1,080°. Viewed from the Earth, Venus has completed 5 plus (1,080° ÷ 360° =) 3, i.e. 8 revolutions with respect to the Zodiac. In the meantime, however, the Earth has also completed 5 revolutions around the sun. At this point, Venus has completed a total of (8 + 5 =) 13 revolutions with respect to the fixed stars.
These same figures should be multiplied by three to correspond to the 24 solar years that make up the Numan cycle.

[5] If this is indeed so, Venus pentagrams on early seventh century Etruscan kraters (see Figs. 9.10 and 9.11) precede by one or two hundred years the attribution to Pythagoras (ca. 580/70-480 BCE) or Parmenides (515-after 450 BCE) of the knowledge that "Hesperus and Lucifer are one and the same" (Diogenes Laertius, *Lives of the Philosophers*, 9.23).

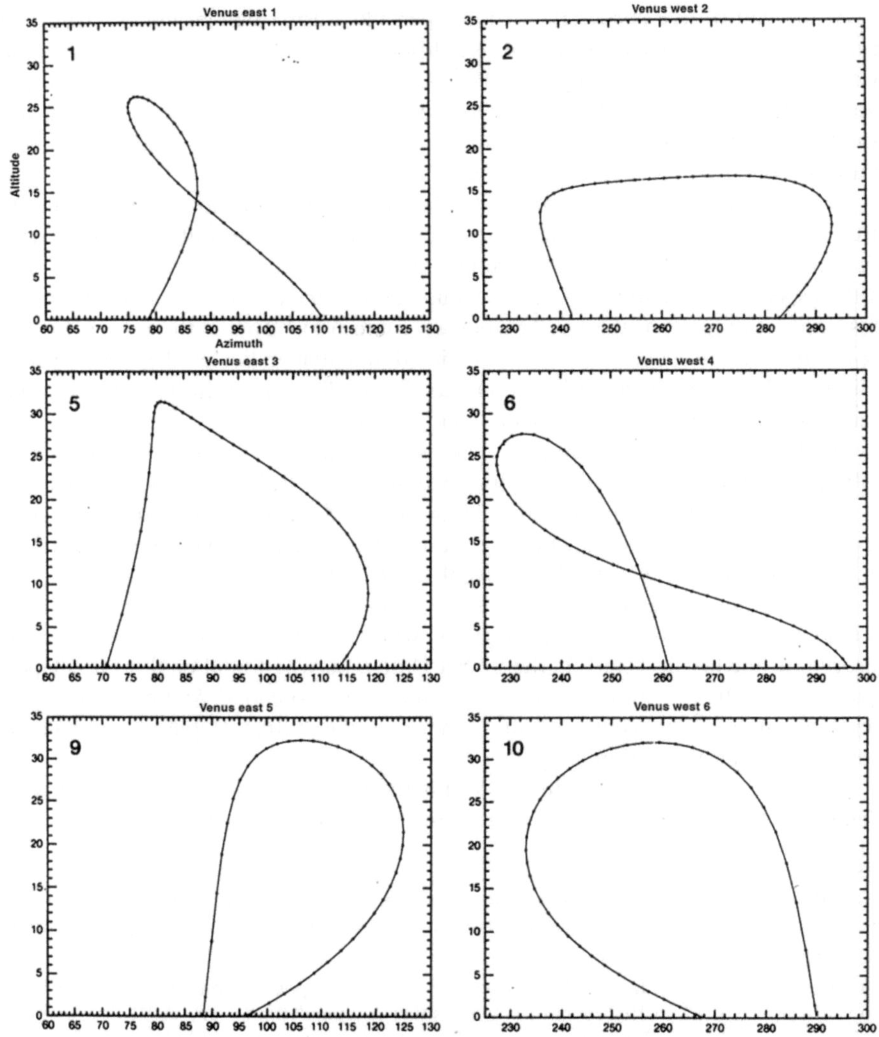

Fig. 9.3 A series of 12 subsequent intervals of Venus's appearance. The planet appears alternately as morning and evening star. Each dot indicates the position with respect to the Sun at 5 day intervals. Panel (*1*) first morning appearance interval; (*2*) first evening appearance interval; (*3*) second morning appearance interval, etc. The pattern is repeated after five pairs of appearance intervals, each lasting around 256 days. Thus, panel 11 repeats panel 1, and so on

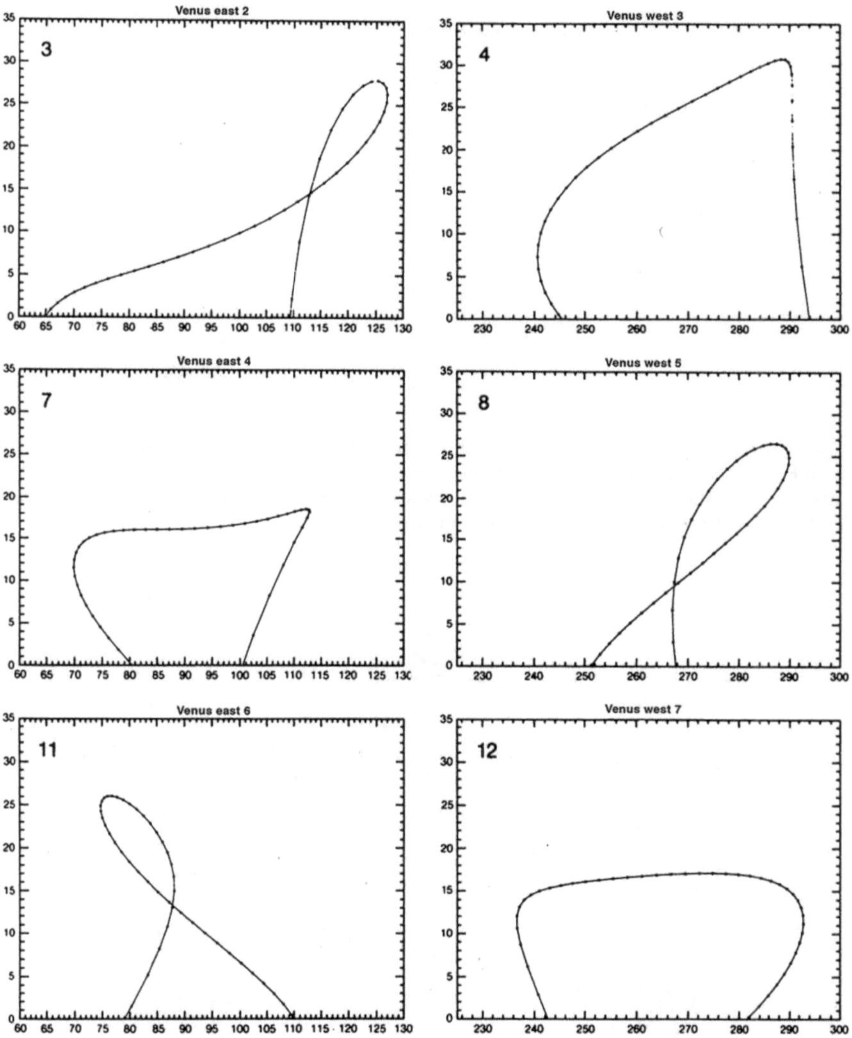

Fig. 9.3 (continued)

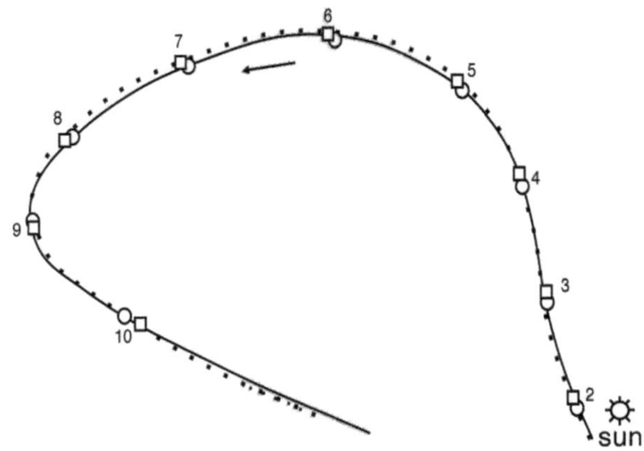

Fig. 9.4 Two subsequent intervals of Venus's appearance as the evening star, 8 years apart. After an 8 year spell, the planet returns almost exactly to its starting position with respect to the Sun and the fixed stars. The numbers 2, 3, etc. indicate the position of the planet on 1 February, 1 March, etc. in 1994 (circles) and in 2002 (squares)

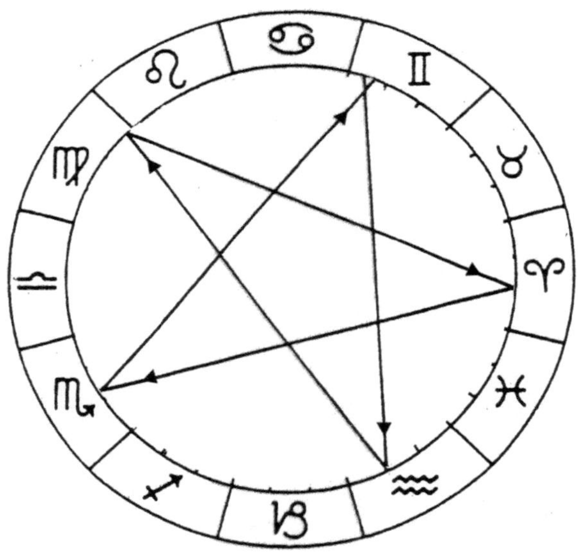

Fig. 9.5 The Venus pentagram. The five points of the star indicate the five Zodiac constellations, and the five moments during the year when, viewed from Earth, the planet Venus returns to the same configurations with respect to the Sun

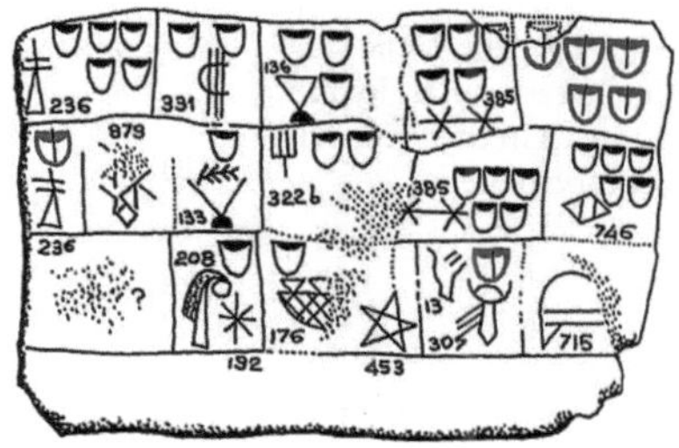

Fig. 9.6 The Venus pentagram in an ideographic Script, Uruk IV, 3500 BCE (from Silvestri et al. 1990)

Fig. 9.7 The Venus pentagram next to a nude female divinity, Anatolia, 2100 BCE. Mould cast in black serpentine, 78 cm in height, provenance (from Bittel 1977) Courtesy: Gallimard, Paris

Fig. 9.8 The Venus pentagram on a Minoan seal, Crete, 1700 BCE

Fig. 9.9 The Venus pentagram on an hydria, Naxos, Greece, XII–XI century BCE

Fig. 9.10 The Venus pentagram in a matrimonial setting (?), Etruria 700–650 BCE. (from Bianchi Bandinelli 1985) Courtesy: Gallimard, Paris

Fig. 9.11 The Venus pentagram in a bellicose setting, Etruria 650 BCE. Crater signed *Aristonothos*, from Cerveteri – Rome, Palazzo dei Conservatori (from Bianchi Bandinelli 1985) Courtesy: Gallimard, Paris

Fig. 9.12 The Venus pentagram on a jewel, Chilikty, Kazakhstan, VII–VI century BCE. (from Kazakhstan 2010) Courtesy: Central State Museum of the Republic of Kazakhstan, Astana

Fig. 9.13 The Venus pentagram on an askos, Lavello, Potenza, Italy, VI century BCE (from Bianchi Bandinelli 1985) Courtesy: Gallimard, Paris

Fig. 9.14 The Venus pentagram and a gold myrtle wrath, Stavroupolis, Thessaloniki, 325–300 BCE. Please note the five-pointed star connected with myrtle, the plant sacred to Venus/Aphrodite. Stavroupolis, Thessaloniki, Greece, 325–300 BCE – Archaelogical Museum of Thessaloniki, Greece (from *The Gold 2008*) Courtesy: DMEEP-Directorate of Museums, Exhibitions and Educational Programs, Athens

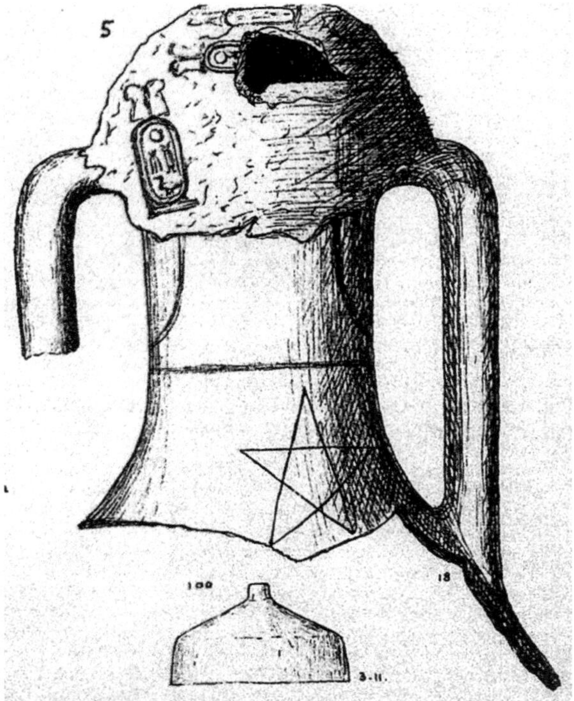

Fig. 9.15 The Venus pentagram on an Chian jar neck, Daphne, Egypt, 550–525 BCE-London, British Museum EA22356

Fig. 9.16 The Venus pentagram on a Roman Hellenistic coin, Pytanè, Greece, c. 12–7 BCE. (from Etcheto 2012) Courtesy: Ausonius Editions/Aquitania, Bordeaux, France

and protected love; the latter, a male deity who promoted and protected war. This counterposition continued in Ancient Rome: Venus/*Hesperus* protected weddings; Venus/*Mater Matuta*/*Lucifer* protected heroes, especially Furius Camillus.

But by far the planet's most important duty was, along with *Fortuna*/the Moon, to protect conception.[6]

[6] For more on Venus/*Hesperus*, protector of weddings, and Venus/*Mater Matuta*/*Lucifer*, protector of conception, see Chaps. 10 and 11.
On *Mater Matuta*, protector of Furius Camillus, see Dumézil 1981, passim, though in any event this author considers *Matuta* to be the Goddess of Dawn; with regard to the identification of *Matuta* with *Lucifer*/Venus at dawn, see Chap. 10 and, at more length, Magini 1996, pp. 34–46, and in particolar Appendix 1, *Furius Camillus, the protégé of Matuta*, pp. 101–104.
For American Indians, Venus/*Lucifer* also looked kindly on warriors; the Sioux nicknamed General Custer the "son of the morning star".

Chapter 10
The Movements of Venus (and the Moon), Female Fertility, and the Feast Days of *Veneralia* and *Matralia*

The motion of Venus is apparent in the rhythm of the many feasts associated with female life and fertility, particularly weddings and conception-related feasts.

At the moment of matrimony, Venus (Fig. 10.1) must shine as the Evening Star. Virgil states:

> 'Mopsus, cut new torches; you are getting married.
> Scatter the nuts, husband: Hesperus leaves Mount Oeta for you'.[1]

Late-period Claudian recalls this when he writes:

> And now its own Evening Star had shone upon the underworld,
> And the maiden is led into the bridal chamber.[2]

Even Catullus, in his most famous wedding song, notes that people were prepared to postpone their wedding to benefit from the heavenly body's splendour and protection:

> BOYS
> 'The Evening Star is here, young men, so rise: the Evening Star
> At last brings heaven its long-awaited bodies of light.
> Now it is time to rise, time to leave the well-stocked tables.
> Now a young maiden will come; the wedding refrain will be sung [...]'
>
> GIRLS
> 'You maidens, do you see the young men? Rise to face them;
> The Night-bringer no doubt reveals Oetean fires [...]'
>
> [...]
> GIRLS
> 'Evening Star, what heavenly body is borne more savagely through the sky?
> You, who can tear a daughter away from her mother's embrace,

[1] Vergil, *Bucolica*, 8.29–30: '*Mopse, novas incidas faces, tibi ducitur uxor; / sparge, marite, nuces, tibi deserit Hesperus Oetam.*' "To leave Mount Oeta" means "the night is falling".

[2] Claudian, *de raptu Proserpinae*, 2.361–2: *Iam suus inferno processerat Hesperus orbi: / ducitur in thalamum virgo.*

L. Magini, *Stars, Myths and Rituals in Etruscan Rome*, Space and Society,
DOI 10.1007/978-3-319-07266-1_10

Fig. 10.1 The nuptials of Heracles and Hebe. In this "recent" depiction, the planet Venus stands in for the goddess. To the right, Aphrodite/Venus, in star-covered robes, oversees the hero's nuptials; in the middle, the veiled bride sits on the bridal bed. Apulian crater, Ceglie, Italy, ca. 350 BCE (from W. H. Roscher, *Ausführliches Lexikon der griechischen und römischen Mythologie*) Author: Wolfgang Rieger—Wikimedia Commons

Tear a daughter away from her mother's embrace though she holds fast,
And give the chaste girl to a young man on fire? […]'

BOYS
'Evening Star, what heavenly body shines more delightfully in the sky?
O you, who strengthen with your flame a wedding pledged,
Which young people and parents have fixed in place beforehand,
They did not join bride and groom before your blaze carried itself away.
What thing more desired do the gods give at a happy hour? […]'

GIRLS
'My friends, Hesperus has taken a maiden away from us […]'.[3]

The most auspicious day for Roman women to marry was 1 April, the date of the *Veneralia*, the feast of *Venus Verticordia* and *Fortuna Virilis*. In the wedding rite—described by Ovid in glorious detail—the Goddess of Love is the ideal prototype for the bride: first, her statue is undressed, then it is washed, dried, dressed again, decorated, and festooned in jewellery by the faithful, who serve as handmaidens, before making an offering of flowers:

[3] Catullus, *Carmina*, 62.1–32: *'Vesper adest, iuvenes, consurgite: Vesper Olympo / exspectata diu vix tandem lumina tollit. / Surgere iam tempus, iam pinguis linquere mensas, / iam veniet virgo, iam dicetur hymenaeus […] '/' Cernitis innuptae, iuvenes? Consurgite contra; / nimirum Oetaeos ostendit Noctifer ignes […] '/' Hespere, quis caelo fertur crudelior ignis? / qui natam possis complexu avellere matris, / complexu matris retinentem avellere natam, / et iuveni ardenti castam donare puellam […] '/' Hespere, quis caelo lucet iucundior ignis? / qui desponsa tua firmes conubia flamma, / quae pepigere viri, pepigerunt ante parentes, / nec iunxere prius quam se tuus extulit ardor. / Quid datur a divis felici optatius hora?[…] '/' Hesperus e nobis, aequales, abstulit unam.'*

Duly do ye worship the goddess, ye Latin mothers and brides
and ye, too, who wear not the fillets and long robe.
Take off the golden necklaces from the marble neck of the goddess;
take off her gauds; the goddess must be washed from top to toe.
Then dry her neck and restore to it her golden necklaces;
now give her other flowers, now give her the fresh-blown rose.

Next, the faithful, now fully qualified as betrothed, undressed for a common group bath—a classical fertility rite—during which incense is offered to Virile Fortune, for whom:

[…] every blemish on the naked body is plain to see;
Virile Fortune undertakes to conceal the blemish and to hide it from the men,
and this she does for the consideration of a little incense.

Finally, the women must make haste and drink

poppy pounded with snowy milk
and liquid honey squeezed from the comb;
when Venus was first escorted to her eager spouse,
she drank that draught: from that time she was a bride.[4]

Ex illo tempore nupta fuit, "from that time she was a bride": repeating every 1 April the gesture once performed by the Goddess, the faithful "re-actualize"—as Eliade would say—the prototypical event fossilized in the words of myth and the gestures of rite: "From that moment onwards" they too become brides (Fig. 10.2), even if it is true that Venus fails to put in an appearance as the Evening Star every single 1 April, but only on 1 April of what were known as "ideal years".

The most auspicious day for conception was 11 June, the day of the *Matralia*—the festival of *Mater Matuta* and *Fortuna*—whose name derives from the *bonae matres*, "women capable of having children" or, better, "women (given) the gift of maternity".

Ovid explains that this day marks the anniversary of the wondrous conception of king Servius Tullius (578–535 BCE), who was to found temples to both goddesses (Fig. 10.3) and decree a common feast day:

There, on this day, it is said, Servius [Tullius] consecrated
with his own sceptered hands a temple to Mother Matuta.[5]

---

[4] Ovid, *Fasti*, 4.133–160; here I quote lines 133–138 and 148–154: *Rite deam colitis, Latiae matresque nurusque / et vos, quis vittae longaque vestis abest. / Aurea marmoreo redimicula demite collo, / demite divitias: tota lavanda est. / Aurea siccato redimicula reddite collo: / nunc alii flores, nunc nova danda rosa est.*; / *[…] et vitium nudi corporis omne videt: / ut tegat hoc celetque viros, Fortuna Virilis / praestat et hoc parvo ture rogata facit. / Nec pigeat tritum niveo cum lacte papaver / sumere et expressis mella liquata favis; / cum primum cupido Venus est deducta marito, / hoc bibit: ex illo tempore nupta fuit.*

[5] Ovid, *Fasti*, 6.479–80: *Hac ibi luce ferunt Matutae sacra parenti / sceptriferas Servi templa dedisse manus.*

The Latin *parens* is the present participle of the verb *pario*, "to beget, to give birth"; the primary meaning is "(she) who is begetting, who is giving birth", from which we have drawn the

Fig. 10.2 The Secundus and Projecta wedding casket. In this "late" depiction, the goddess stands in for the planet Venus. On the lid, Venus decks herself out while looking in the mirror; lower down and in the centre, the seated bride imitates the goddess's movement and adorns herself, assisted by two handmaidens. The twin peacocks recall Juno, while the doves connote Venus—both of them are protector goddesses of matrimony. Silver, 28 × 55 cm, from the Esquiline, late fourth century CE—London, British Museum (from Bianchi Bandinelli 1970) Courtesy: Gallimard, Paris

Tullius' mother, Ocresia, a slave to Queen Tanaquilla, was made pregnant by the flames in the fireplace, that is to say by the God Vulcan, who was the true father of the servant destined to become the king of Rome. This is how the poet recounts the tale:

> For the father of [Servius] Tullius was Vulcan,
> his mother was the beautiful Ocresia of Corniculum.
> After performing with her the due sacred rites, Tanaquil
> ordered Ocresia to pour wine on the hearth, which had been adorned.
> There among the ashes there was, or seemed to be, the shape
> of the male organ; but rather the shape was really there.
> Ordered by her mistress, the captive Ocresia sat down at the hearth.
> She conceived Servius, who thus was begotten of seed from heaven.[6]

(Footnote 5 continued)
more general meaning of "mother, father, parent". So, the literal meaning of *Matutae [...] parenti* is "to the expectant Matuta".

[6] Ovid, *Fasti*, 6.473–636. Quoted here are lines 627–634: *Namque pater Tulli Volcanus, Ocresia mater / praesignis facie Corniculana fuit. / Hanc secum Tanaquil sacris de more peractis / iussit in ornatum fundere vina focum: / hic inter cineres obsceni forma virilis / aut fuit aut visa est, sed fuit illa magis. / Iussa foco captiva sedet: conceptus ab illa / Servius a caelo semina gentis habet.* Note that the Latin *praesignis* does not mean "beautiful" but "foreboding".

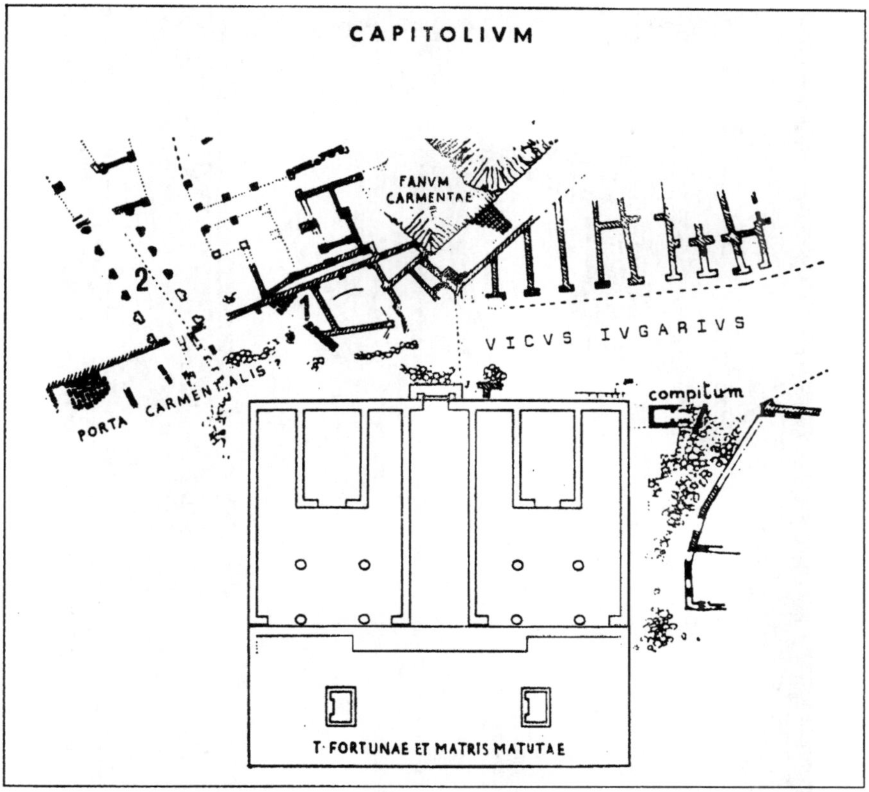

Fig. 10.3 The twin temples of *Fortuna* and *Mater Matuta*: a Republican era reconstruction. Founded by Servius Tullius, this double sanctuary lies at the southernmost portion of the base of the Capitol, near *Porta Carmentalis*. (from Coarelli 1988)

At both of these feasts, from their places in the heavens two feminine deities watch over and protect the faithful on Earth. At the Calends of April, the goddesses are *Venus Verticordia* and *Fortuna Virilis*; on 11 June, *Mater Matuta* and *Fortuna.*

Fortuna is represented in the sky by the Moon (see Chap. 8 and Fig. 8.1), a heavenly body that is no less variable and changeable than the Goddess.

Venus and *Mater Matuta* (Fig. 10.4) are both represented by the planet Venus.

In the most propitious of years, on 1 April, Venus is *Hesperus*, the Evening Star; on 11 June, *Mater Matuta* is *Lucifer*, the Morning Star. Ovid opens his commentary on the feast of *Matralia* thus:

Fig. 10.4 Mater Matuta. The conventional name of this female divinity underscores her association with fertility and procreation; here, this attribute is highlighted by the infant in her lap. Pietra fetida, Chianciano, ca. 450 BCE—Museo Archeologico, Chianciano, Italy. Author: Edisonblus—Wikimedia Commons

[…] and the watchful Morning Star comes forth the eastern waters.[7]

As *matutinus*, he is the natural opposite of *vespertinus*; *Matuta* is the reversed image of *Vesper.*

In the Numan calendar, 71 days elapsed between the Calends of April and 11 June—29 days in April, 31 in May and 11 in June: counting inclusively, 29 + 31 + 11 = 71 days. This figure corresponds to the 71 days that the planet Venus takes to travel from lower conjunction with the Sun to its greatest western elongation (see Chap. 9 and Fig. 9.1). In the first of these positions, viewed from the Earth, the planet appears to be superimposed over the Sun, and is therefore invisible; in the second position, the planet lies at its greatest western elongation from the Sun, equivalent to around 47°, and rises over the horizon, as the Morning Star, around 3 h before the Sun.

Concordances between the movements of the planet Venus in the sky and feast days on Earth for the women of Rome don't end there:

- 75 days[8] before the Calends of April is the feast of *Carmenta Postvorta*, on 15 January of the preceding year, when the planet is at its greatest eastern elongation from the Sun, visible as the Evening Star for up to 3 h after the Sun sets;
- 222 days[9] before that is the feast of *Carna* on the Calends of June, when the planet is in higher conjunction with the Sun, and is not visible from Earth;
- 585 days[10] after 11 June—when the year is intercalated—is the feast of *Carmenta Antevorta* on 11 January of the following year, when Venus completes a synodic cycle and once more reaches its greatest western elongation with respect to the Sun, and rises over the horizon as the Morning Star around 3 h before the Sun (see Table 10.1).

If we take conception the day of the *Matralia*, on 11 June, as our starting point, on 13 January—the mid-point between the two *Carmenta* feast days—is the 210th day,[11] corresponding to the 7th month of pregnancy; this is the moment that the unborn child begins to turn around in the mother's womb and assume the most natural and least perilous position for birth.

The two *Carmentas*, *Antevorta* (or *Prorsa*) and *Postvorta*, lie in relationship to the two positions of the foetus, when either the feet or the head are facing down. Gellius, developing what Varro had to say, makes the following unequivocal statement:

---

[7] Ovid, *Fasti*, 6.474: *[…] et vigil Eois Lucifer exit aquis.*

[8] In Roman times, the practice was to count inclusively: 15 days in January + 28 in February + 31 in March + 1 in April = 75 days.

[9] 29 days in June + 178 in July, August, September, October, November and December + 15 in January = 222 days.

[10] 19 days in June + 257 days up to 28 February = 276; 276 + 309 days from 1 March to 11 January = 585 days.

[11] 19 days in June + 178 in July, August, September, October, November and December + 13 in January = 210.

Table 10.1 The movements of Venus and female fertility festivities

Interval in days	Date of intercalary cycle	Festivity and/or deity celebrated	Configuration of Venus in its cycle	Female fertility landmarks
222	1 June Year Minus One	Calendae Fabariae/ *Carna*	Higher conjunction Venus invisible	
75	15 January Year Minus One	CARMENTALIA/ *Carmenta Postvorta*	Greatest eastern elongation Venus visible as Evening Star	7th month of pregnancy
	1 April Year Zero* (intercalated)	Veneralia/ *Venus Verticordia* and *Fortuna Virilis*	Lower conjunction Venus invisible	Wedding feast
71	11 June Year Zero	MATRALIA/ *Mater Matuta* and *Fortuna*	Greatest western elongation Venus visible as Morning Star	Feast of conception
585	11 January Year Plus One	CARMENTALIA/ *Carmenta Antevorta*	Greatest western elongation Venus visible as Morning Star	7th month of pregnancy

*The identification of Year Zero—the "ideal year" in the Numan cycle—during which Venus makes an apparition at weddings and conceptions in the ideal configuration should be the 16th year, according to the data presented in Chap. 7. But this is yet to be confirmed

> When therefore, contrary to nature, they [babies in the womb; author's note] are turned upon their feet, and retained in the womb, with their arms extended, women are delivered with great difficulty.
> For the purpose of averting this calamity, altars were erected in Rome at the two temples of the Carmenta, one of which was called *Postvorta* and the other *Prorsa*, and which protect and take their names from the different births, natural and contrary to nature.[12]

Censorinus, another valuable source, notes that, according to Chaldean/Pythagorean doctrine, this length of time coincides with the earliest possible date that a foetus may be born:

> Ready in less time, the seven-month child leaves the womb 210 days after conception.[13]

---

[12] Aulus Gellius, *Noctes Atticae*, 16.16: *Quando igitur contra naturam forte conversi in pedes brachiis plerumque diductis retineri solent aegriusque tunc mulieres enituntur, huius pericoli deprecandi gratia arae statutae sunt Romae duabus Carmentibus, quarum altera 'Postvorta' cognominatast, 'Prorsa' altera a recti perversique partus et potestate et nomine.*

[13] Censorinus, *de die natali*, 11.2: *[…] alterum minorem, quem vocant septemmestrem, qui decimetducentesimo die post conceptionem exeat ab utero.* See also 9.3, 11.5 and 11.11.

Chapter 11
The Length of Pregnancy, the Solar New Year, and the Feasts of *Matronalia* and *Liberalia*

Censorinus informs us that, according to the same Chaldean/Pythagorean doctrine, the standard length of a pregnancy is associated with astronomical data. It takes 273 point something days, or ¾ of the solar year, to travel from conception to birth:

> Given that the Sun circumnavigates the Zodiac in 365 days and a handful of hours, if we take away a fourth, that is to say 91 days and a few hours, we may calculate that the Sun travels three quarters of the way round its orbit in a little less than 274 days, until it reaches a position from which it looks towards the point of conception in quadrature.

According to the same Censorinus, Pythagoras points out that this is an odd, and therefore an auspicious number

> as it comprises an odd number of whole days plus, to complete the period, a portion of the following day which, it should be noted, is still less than a whole day.[1]

It follows that if the ideal conception takes place on 11 June, on the feast of the *Matralia*, the ideal day to be born coincides with the solar new year on 9 March (see Chap. 7), 273.25 days[2] later.

---

[1] Censorinus, *de die natali*, 11.9 and 11.11: *cum signiferum orbem diebus CCCLXV et aliquot horis sol circumeat, quarta necesse est parte dempta, id est diebus LXXXXI aliquotque horis, tres quadras reliquis diebus CCLXXIIII non plenis percurrat, usque dum perveniat ad id loci unde conceptionis initium quadratus aspiciat; […] enim impares […] dicit expleri, ad quorum consummationem aliquid ex sequentibus accedere, quod tamen diem solidum non adferat.*

See also Aulus Gellius, *Noctes Atticae*, 3.10.8: "[…] babies […] come into this world two hundred and seventy three days after conception, that is to say, after the fortieth week has commenced, *[…] hi […] post ducentos septuaginta tres dies, postquam sunt concepti, quadragesima denique hebdomade inita nascuntur.*"

[2] Naturally, we must also consider the average of 10.25 intercalary days per year in the Numan cycle (see Chap. 4). The resulting calculation is: 19 days in June + 235 in July, August, September, October, November, December, January and February + 10.25 intercalary days + 9 in March = 273.25.

L. Magini, *Stars, Myths and Rituals in Etruscan Rome*, Space and Society,
DOI 10.1007/978-3-319-07266-1_11

In Rome, no public feast was held on 9 March; the birth of a common citizen remained a private event. But two feasts were held one *nundina* before and one *nundina* after 9 March.[3]

One *nundina* before the date of birth, on 1 March, pregnancy completes the 9th and enters the 10th lunation. As Macrobius explains, this is the moment when "in the uterus the fertilized seed ripens into life in the 10th month, retained by delicate natural bonds until it issues into the light."[4] This is the moment when, approaching the start of labour and all of the attendant dangers for the mother and newborn infant, we have the feast of *Matronalia*, which kicks off the lunar year at the Calends of the first month of the year. This particular feast is for *matronae*:

> *Matrona* is the name given to a woman married to a husband, provided that thus she remains, even if she has no children; she is so called from the word *mater*, referring to the fact that, though she has not yet become a 'mother', she has the hope and bears the promise of becoming one soon.[5]

On this day, Latin mothers—"for in their travail they both fight and pray"—journeyed to the *Iuno Lucina* temple on the Esquiline Hill and invoked the goddess with the words:

> 'Thou, *Lucina*, hast bestowed on us the light (*lucem*) of life;
> [...]'thou dost hear the prayer of women in travail.'

By 1 March a *matrona*, who had conceived on the preceding 11 June and has completed the 9th lunation of pregnancy would be preparing to give birth a few days later, on the 9th of the month. On the verge of realizing "the hope and promise of motherhood", she follows the instructions chronicled by the poet:

---

[3] The *nundinae* are an Etruscan institution that was carried over into the Roman calendar: an eight-day cycle, it consists of seven days tilling the fields and one day in the town market. These eight days are marked by the *nundinal* letters A to H. Counting inclusively, from one day A to the next makes a nine-day cycle, hence the name *nundinae*. New Year's day was always considered an "A" day.

In the singular, *Nundina* means "market day". *Nundina* is also the name of the *dies lustricus* deity (see Macrobius, *Saturnalia*, 1.16.36): "For the Romans, among other things *Nundina* stood for the name of a goddess so called for the 9th day of life in newborns, known as the lustral day. The lustral day [...] is the 9th day for males., *Est etiam Nundina Romanorum dea a nono die nascentium nuncupata, qui lustricus dicitur. Est autem dies lustricus [...] maribus nonus.*"

Perhaps instituted by Servius Tullius (Macrobius, *Saturnalia*, 1.16.33), the concept of the *nundina* has survived to the present day in the religious rite of 'novena'.

[4] $9 \times 29.5306 = 265.7754$; 19 days in June + 235 in July, August, September, October, November, December, January and February + 10,25 intercalary days + 1 day in March = 265.25.

[5] Aulus Gellius, *Noctes Atticae*, 18.6.8: *[...] matronam dictam esse proprie, quae in matrimonium cum viro convenisset, quoad in eo matrimonio maneret, etiamsi liberi nondum nati forent, dictamque ita esse a matri nomine, non adepto iam, sed cum spe et omine mox adipiscendi.*

> [...] let her who is with child unbind her hair before she prays,
> in order that the goddess may gently unbind her teeming womb.[6]

One *nundina* after the date of birth, on 17 March, is the moment that males born on 9 March undergo purification and naming rituals.[7] This *dies lustricus*, "purification day", coincides with the feast of *Liberalia* on 17 March, when the birthday of those born at the same time and in the same age group (see Chap. 7) is celebrated. The feast of *Liberalia*—the day on which males reach the age of majority—should also be regarded as an anniversary of the *dies lustricus* of those born as part of the same age group.

The *Matronalia* on 1 March, the solar New Year's day on 9 March and the *Liberalia* on 17 March—all of which are feasts in preparation for labour, birth, and name-giving for male newborn babies—are nine days' apart, and therefore all have the same *nundinal* letter A.

Table 11.1 summarizes what we have seen thus far (and, moreover, what we shall see later in Chaps. 14, 15 and 24) regarding the relationship between the motion of heavenly bodies and life landmarks on Earth, as marked by festivities throughout the Numan year.

Once again, we find the same inevitable intermingling and harmonious integration of life landmarks for the various generations:

- Initiation rites for adolescents (male and female) commence on 1 October in Year X, during the feast of *Tigillum sororium*,[8] and conclude (for females) on 15 March of the following year, at the feast of *Anna Perenna* (see Chap. 14);

---

[6] Ovid, *Fasti*, 3.243–58; quoted here are lines 244 and 255–258: *[...] quarum militiam votaque partus habet; [...] 'Tu nobis lucem, Lucina, dedisti. / [...] tu voto parturientis ades!' / Si qua tamen gravida est, resoluto crine precetur / ut solvat partus molliter illa suos.*

The terms "fight" and "pray" hark back to Mars, Rhea Silvia and the conception of Romulus, the ideal prototype for the conception of a Roman citizen.

If we look at the calendar, we find that on the Calends of March the Moon should be a crescent high in the sky at sunset, 1.5 days after new Moon; this would make it impossible to refer to the goddess that the heavenly body represents as *Lucina*, "she who brings the light, giver of light". And yet on those very years when the solar New Year and new Moon fall on 8 or 9 March (see Chap. 7), on the preceding Calends the Moon was in its final quarter, high in the sky at sunrise and heralding the Sun's arrival. The Latin name *Lucina* may be translated into Greek as *Phosphoros* (as it was translated by Dionysius of Halicarnassus in his book *Roman Antiquities* 4.15.5), i.e. "bringer of light" – and, once again, akin to Venus/Lucifer.

[7] This reconstruction of ancient Rome bears a striking correspondence to the traditions handed down to us in the Christian calendar: 1) Annunciation of the Virgin, preceding conception (Luke 1.31 and 2.21), which is celebrated on 25 March; 2) A child is born nine months later (now equivalent to 275 days) on the night of 24/25 December; 3) According to Jewish tradition, eight days after birth, the child is taken to the temple to be given its name and for circumcision, which brings us to 1 January, our modern day New Year's Day. It is worth noting that, until recent times, in Florence 25 March was considered the first day of the year: it was celebrated along with the conception of Christ, the Spring equinox and New Year's Day.

[8] On initiation rites and *Tigillum sororium*, see Torelli 1984, p. 106; with references to A. von Domaszewski, *Abhandlungen zur roemischen Religion*, Leipzig-Berlin 1909, p. 222 and K. Latte, *Roemische Religiongeschichte*, Muenchen 1960, p. 97 and 133.

Table 11.1 Roman feasts, life landmarks and the motion of heavenly bodies

Day, feast day and deity of the Numan year	Life landmark	Movement of celestial bodies	Topic covered in:
1 OCTOBER *Tigillo sororio*	Beginning of female and male initiation	28/29 September: Anniversary of the 16, 35 and 37th lunar eclipses	Chap. 15. Table 15.2
15 MARCH *Annae Perennae*	Conclusion of female initiation: last menstruation before marriage	Zodiac new year Anniversary of the 43, 47 and 49th solar eclipses	Chaps. 24, 14 and 15. Table 15.1
17 MARCH LIBERALIA	Conclusion of male initiation: *Liberi*, or "free men", come of age	Last day of the solar year (non-intercalated years)	Chaps. 7 and 11
1 APRIL Veneralia/*Venus* and *Fortuna Virilis*	WEDDING FEAST	Year X:* Venus in lower conjunction	Chap. 10. Table 10.1
11 JUNE MATRALIA/ *Mater Matuta* and *Fortuna*	FEAST OF CONCEPTION: 1ST DAY OF PREGNANCY	Year X: Venus at greatest western elongation	Chap. 10. Table 10.1
11–15 JANUARY CARMENTALIA/ *Carmenta Antevorta* or *Carmenta Postvorta*	210TH DAY OF PREGNANCY: FOETUS TURNS IN THE WOMB AND/OR BIRTH OF SEVENTH-MONTH BABIES	11 Jan. Year X Plus One: Venus at greatest western elongation 15 Jan. Year X Minus One: Venus at greatest eastern elongation	Chap. 10. Table 10.1
1 MARCH MATRONALIA / *Iunoni Lucinae*	END OF THE 9TH LUNATION OF PREGNANCY: FEAST OF PREPARATION FOR LABOUR	Lunar new year Anniversary of the first solar eclipse	Chaps. 11 and 14. Table 15.1
9 MARCH	FEAST OF BIRTH-GIVING: 273RD DAY OF PREGNANCY	Solar new year 8 March: anniv. of the 17 and 19th lunar eclipses	Chaps. 7, 11 and 14. Table 15.2
17 MARCH LIBERALIA	ANNIVERSARY OF THE *DIES LUSTRICUS* FOR MALES	Last day of the solar year (non-intercalated years)	Chaps. 7 and 11

*The term Year X refers to an "ideal year" in this cycle—which as yet requires identification—during which the planet Venus is to be found in what is considered its ideal position

- Young women are consequently ready for marriage on 1 April, in time for the feast of *Veneralia*;
- Women who marry on 1 April in Year X and conceive on 11 June on the feast of *Matralia* complete their 210th day of pregnancy on 13 January on the feast of *Carmentalia*, prepare for birth on 1 March the following year, complete their ninth lunation of the pregnancy in correspondence with the feast of *Matronalia*,

and give birth on 9 March, which is the solar New Year's Day; the *dies lustricus* for their newborn boys falls on 17 March, the feast of *Liberalia.*

The temporal differentiation between these two aspects of female life corresponds closely to the climatic and biological laws of nature:

- almost six full cycles of the Moon are dedicated to initiation—from 1 October to 15 March, covering autumn and winter;
- 71 days are reserved for matrimony and conception (the number of days in the "ideal" year it takes Venus to transit from lower conjunction with the Sun to its greatest western elongation), corresponding to 2½ cycles of the Moon from 1 April to 11 June, at the heart of spring;
- pregnancy corresponds to something more than nine lunations[9] from 11 June to 9 March, taking in summer, autumn and winter.

With so many feast days associated with movements in the heavens, the calendar provided a rhythm for the life of Roman women during an entire ideal year.

---

[9] The ninth lunation—or synodical month—runs from the 266th to the 295th day after conception: birth occurs on the 273rd day, in the tenth lunation (once again, counting inclusively).

See Ovid, *Fasti*, 3.124: "a woman brings forth in twice 5 months, *bis quinto femina mense parit.*", and *Fasti*, 2.175: "ten times the horned moon had filled her orb afresh, *luna novum decies implerat cornibus orbem.*"

See also Aulus Gellius, *Noctes Atticae*, 3.10.8, quoted in note 11.1 above, who establishes birth "two hundred and seventy three days after conception, that is to say, after the fortieth week has commenced".

Chapter 12
The Solstice Feasts: *Fors Fortuna* and *Saturnalia*

The Roman calendar had a very large number of festivities scattered throughout the year. Two of these festivities may, for simplicity's sake, be characterized as Rome's solstice celebrations:

- the first festival, which by tradition was established by King Servius Tullius (578–533 BCE), was dedicated to *Fors Fortuna*, between 24 and 26 June;
- the second festival, which, always according to tradition, was even older and predated the foundation of the city itself, was the *Saturnalia*, which ran between 17 and 23 December.[1] The first of these festivities fell on the summer solstice; the second on the winter solstice.[2]

To recap: the festival of *Fors Fortuna* occurs around the summer solstice, during the longest days of the year, when the sun is in the sign of Cancer; the festival of *Saturnalia* straddles the period around the winter solstice, during the shortest days of the year, when the sun enters or is about to enter the sign of Capricorn.[3]

---

[1] On the goddess *Fortuna* as represented in the sky by the new Moon, see Chap. 8; see also Magini 1996, pp. 32–34; Magini 2006, pp. 123–134; Magini 2008, pp. 184–188.
For a more detailed examination of the feasts of *Fors Fortuna* and the *Saturnalia*, see Magini 2008, pp. 207–214.

[2] Different authors place the summer solstice at different times: for Ovid, *Fasti*, 6.790, it falls on 26 June; for Columella, *de re rustica*, 11.2.48, between 24 and 26 June; for Pliny, *Naturalis Historia*, 18.256, 264 and 288, on 24 June.
As for the winter solstice, according to tradition (see Chap. 26)—notably, Varro, *de lingua latina*, 6.8, Ovid, *Fasti*, 1.163–4, and the *Praenestine Calendar*—it falls on 21 December; for Pliny, *Naturalis Historia*, 18.221, it is *fere*, "around", the 25.

[3] As for the Sun's entry into Cancer, as far as Ovid, *Fasti*, 6.727, and Columella, *de re rustica*, 11.2.48, are concerned, it takes place on 19 June. Pliny, *Naturalis Historia* 18.221, limits himself to writing "*bruma capricorni [...] solstitium cancri*, the winter solstice (occurs) in Capricorn [...] the summer solstice in Cancer".

For Columella, *de re rustica*, 11.2.94, the Sun enters the sign of Capricorn on 17 December, the first day of the *Saturnalia*.

L. Magini, *Stars, Myths and Rituals in Etruscan Rome*, Space and Society,
DOI 10.1007/978-3-319-07266-1_12

To the Ancients, Cancer and Capricorn were the signs that linked the microcosm to the macrocosm. In the microcosm, every man lives out his own mortal life; in the macrocosm, the souls of human beings enjoy their immortal existence. The former is on this lowly Earth; the latter is on high in the heavens—or even on the Moon. The two signs of Cancer and Capricorn facilitate the perpetual cycle of the souls, allowing them to come down from the afterlife to Earth, and to rise up from Earth to the afterlife.

To be more precise: souls descend to Earth from birth through the gates of Cancer (or Boreas); souls ascend to the heavens after death through the gates of Capricorn (or Notus).[4]

This conception was expressed in famous lines by Homer, on which Porphyry commented in his *On the Cave of the Nymphs*:

> Perpetual waters through the grotto glide, a lofty gate unfolds on either side;
> to the north, to Boreas, is the way down to mankind:
> the other, to the sacred south, to Notus, is for the gods
> and may not be passed by man, as it is the path of the immortals.[5]

Interestingly, the conception described by Homer (ca. 8th century BCE) would appear to pre-date the true 'invention' of the Zodiac—or perhaps it would be more accurate to say 'codification' in its final version of 12 signs, each of which spans 30°—which scholars believe occurred in Mesopotamia sometime "around the middle of the fifth century BCE".[6]

Two elements are worthy of note: one astronomic, the other more closely associated with Roman festivities. The astronomical element concerns the role played by the Moon and Saturn. In reality, these two heavenly bodies are "errant" and opposite bodies, inasmuch as: the Moon is the body that lies closest to the Earth and the farthest from the fixed stars; Saturn is the farthest from the Earth and the closest to the fixed stars.

Moreover, these two heavenly bodies share a series of correspondences that will not have eluded the earliest sky watchers:

- the Moon takes roughly 29½ days to complete its monthly cycle and return to its initial position with respect to the Sun;
- Saturn takes roughly 29½ years to complete its orbit and return to its initial position with respect to the fixed stars;
- over the 29½ or so years that it takes Saturn to complete its orbit and return to its initial position with respect to the fixed stars, the Moon completes around

---

[4] As with the tropics, the entrance to Cancer—or Boreas—faces north, and the entrance to Capricorn—or Notus—faces south.

[5] Homer, *Odyssey*, 13.109–12: *εν δ υδατο αεναοντα. δυω δε τε οι θυραι εισιν, / αι μεν προς βορεαο καταιβαται ανθρωποισιν, / αι δ αυ προς νοτου εισι θεωτεραι : ουδε τι κεινη / ανδρες εσερχονται, αλλ αθανατων οδος εστιν.*

[6] Britton-Walker 1996, p. 49.

365 29½-day cycles, at the end of which it returns to its initial position with respect to the Sun;

- during these same 29½ years, the Sun self-evidently completes around 29½ of its annual 365-day cycles, and passes that same number of times through its initial position with respect to the fixed stars.

What this means is that even the earliest sky watchers will have noticed the real and gradually increasing distance of these three bodies from the Earth: the Moon, the Sun and Saturn. They will have also have begun to glimpse the ideal and secret harmony of the celestial spheres that would find full expression first with Pythagoras (ca. 570–495 BCE), then with Kepler (1571–1630 CE).

The other bond that associates the Moon with the Tropic of Cancer and Saturn with the Tropic of Capricorn is solely the product of the imagination of the Ancients—an act of imagination that, in any event, starts off with a kernel of reality. Once again, we hear from Porphyry:

> There are two extremities in the heaven: the winter tropic, than which nothing is more southern, and the summer tropic, than which nothing is more northern. The summer tropic is in Cancer, and the winter tropic in Capricorn.
> Since Cancer is nearest to us, it is very properly attributed to the Moon, which is the nearest of all the heavenly bodies to the earth. As the southern pole by its great distance is invisible to us, hence Capricorn is attributed to [Saturn,] the highest and most remote of all the planets.[7]

Though its origins may have been lost in the mists of time, astrological doctrine has retained a similar conception of things ever since: the Moon has its domicile in Cancer and its exile in Capricorn; Saturn has its domicile in Capricorn and its exile in Cancer.

The result of all this is a long sequence of oppositions:

- the first opposition is of the "nearby" Moon compared to the "distant" Saturn;
- the second opposition has the Tropic of Cancer, which is also considered to be "near", in opposition to the Tropic of Capricorn, which is felt to be "distant".
- The third opposition—which we have already touched upon—is between the solstices:
- the summer solstice takes place during the longest days of the year;
- the winter solstice occurs during the shortest days of the year.

Most importantly, if the Sun is in the sign of Cancer, then the Moon—if it is new[8]—will be in the same sign of Cancer, that is to say, at its closest point to the

---

[7] Porphyry, *The Cave of the Nymphs*, 21: *[…] δυο ειναι εν ουρανω ακρα, ων ουτε νοτιωτερον εστι του χειμερινου τροπικου ουτε βορειοτερον του θερινου. εστι δ ο μεν θερινος κατα καρκινον, ο δε χειμερινος κατ αιγοκερων. και προσγειοτατος Σεληνη απεδοθη, αφανους δ ετι οντος του νοτιου πολου τω μακραν ετι αφεστηκοτι και ανωτατω των πλανωμενων παντων ο αιγοκερως απεδοθη.*

[8] The lunar cycle and the solar cycle are not in synch. The moon is new on 24 June once every 19 years; before and after this, it is in a different phase. The mechanism is the same one that

Earth. This means that if the Moon is new during the summer solstice, it will be the closest heavenly body to the Earth and, at the same time, be passing through the point that—ideally—is closest to the Earth. The new Moon happens to be represented in the heavens by the goddess *Fortuna* (see Chap. 8), who on the days between 24 and 26 June is celebrated in the most variable and unpredictable of events: *Fors Fortunae*, the 'Destiny of Fortune'.

It is as if the opposition between the Moon—most particularly the new Moon—and Saturn exalts, and indeed acquires, its greatest value at the two solstices. The Ancients considered the two heavenly bodies to play opposite roles:

- at the summer solstice, when the Sun is in Cancer—that is to say, in the domicile of the Moon—the new Moon (the planet closest to the Earth) is also in Cancer, and is therefore passing through the point closest to the Earth, which favours the descent of souls to the Earth;
- at the winter solstice, when the Sun is in Capricorn—that is to say, in the domicile of Saturn—it is Saturn, the planet farthest from the Earth and closest to the fixed stars, that promotes the re-ascendance of those same souls from the Earth to the heavens.

The second element of interest—one which is more properly associated with Roman feasts—casts light on the meaning behind the rites observed:

- on the feast of *Fors Fortuna*, during the night of the summer solstice, people used to float down Rome's river in boats by night, as Ovid depicts:

> Come, Quirites, celebrate with joy the goddess Fors!
> On Tiber's bank she has her royal foundations.
> Speed some of you on feet, and some in the swift boat,
> and think no shame to return tipsy home from your ramble.
> Yeflower-crowned skiffs, bear bands of youthful revelers,
> and let them quaff deep draughts of wine on the bosom of the stream[9];

- this *Tiberina descensio* signified a journey towards the afterlife—in Porphyry's words, a moving "[…] on the waters […] for souls descending into generation fly to moisture"[10];
- for the feast of the *Saturnalia*, at the winter solstice, slaves donned the clothing of free men and everybody gave one another a gift; this behaviour meant that

(Footnote 8 continued)
regulates the cadence of Easter for Christianity, which falls on the first Sunday following the full moon after the spring equinox.

[9] Ovid, *Fasti*, 6.775–80: *Ite, deam laeti Fortem celebrate, Quirites: / in Tiberis ripa munera regis habet. / Pars pede, pars etiam celeri decurrite cumba, / nec pudeat potos inde redire domum. / Ferte coronatae iuvenum convivia lintres, / multaque per medias vina bibantur aquas.*

[10] Porphyry, *The Cave of the Nymphs*, 10: *[…] τας των υδατων προεστωσας δυναμεις ιδιως […] τας εις γενεσιν κατιουσας ψυχας κοινως απασας.*

> those who are now born slaves will be liberated through the *Saturnalia* festival, and the house attributed to Saturn [i.e. Capricorn], when they live again and return to the fountain of life.[11]

So, to recap anew:

- on 24–26 June, for the festival of *Fors Fortuna*, the sun is in Cancer, the domicile of the Moon. If the Moon is new, it too will be in Cancer, and souls will find their descent to Earth facilitated;
- on 17–23 December, for the festival of the *Saturnalia*, the Sun is in Capricorn, the domicile of Saturn, and souls will find their re-ascent to the heavens facilitated.

The first thing to note is that these theoretical conceptions may hail from Greece, but their practical application takes place in Rome. However, the divinities involved—*Fors Fortuna* and *Saturnus*—are not Roman: the focal point of the cult of *Fortuna* was in the Etruscan city of Praeneste (modern day Palestrina), and was brought to Rome by Etruscan king Servius Tullius; the name *Saturnus* "may be of Etruscan origin".[12] This case repeats what we will see on other occasions[13]: though the documentation available to us is Roman, in all likelihood its origins lie in the East, and arrived in Rome through Etruscan intercession.

A second observation. Since time immemorial, man has considered the summer solstice to be one of two times during the year in which it is easiest to find confirmation of the absolute, predictable and given reliability of the Sun's motion. Year after year, the brightest body in the sky always rises and sets at the same time, and at the same azimuth. The reason for establishing the feast of a lunar divinity such as Fortuna—specifically, the festivity of *Fors Fortunae*, 'the Destiny of Fortune'—on this day probably signifies an attempt to find a way to answer the ancient question: how, when and where will the Moon rise and set on the summer solstice? In other words, compared to the self-evident and immutable behaviour of the "male" Sun, how will the most unpredictable and "feminine" of heavenly bodies behave? What face will the celestial body defined as a kind of "fortune squared" show? What is the "destiny", the "fate", the "fortune" of the *Fortuna*/ Moon on the solstice? Clearly, it took the contemporaneous observation of the Sun and Moon over a very long period indeed to identify the cycle that we know today as the Metonic cycle, in which, every 19 solar years, the Sun and the Moon return almost exactly to the same point in the Zodiac on the same day of the year.[14]

---

[11] Porphyry, *The Cave of the Nymphs*, 23: *[…] οι νυν οντες δια την γενεσιν δουλοι δια της Κρονικης εορτης και του ανακειμενου Κρονω οικου ελευθερουνται, αναβιωσκομενοι και εις απογενεσιν απερχομενοι.*

[12] Ernout-Meillet 1979, s.v.

[13] For example, as regards the feasts of *Regifugium* (24 February) and of the *Vestalia* (9 June), see Chap. 16. For a more detailed explanation, please see Magini 2001, pp. 93–104, and Magini 2005, pp. 1–5.

[14] 19 solar years of 365.2422 days make a total of 6,939.6018; 235 lunations of 29.5306 days total 6,939.6910 days.

This cycle was utilized in Babylon for calendars no earlier than the beginning of the fifth century BCE,[15] before arriving in Athens and receiving the name of the Greek astronomer in 432. However, according to Diodorus Siculus,[16] it was already well known by the "hyperborean" men of Stonehenge...

A third and final observation. The idea of that other *Fortunae sors*, literally the "Lot of Fortune", which truly has had good fortune in later astrological conceptions, must have arisen from the contemporaneous observation of the Sun and the Moon, and from the incessant and unpredictable modification of their angular distance. This procedure consists of measuring the angular distance between the Sun and the Moon at the moment when a given individual is born, and then travelling the same distance from the ascendant in an anticlockwise direction. Such a calculation identifies the "Lot of Fortune", which can then be used to draw up a veritable "lunar horoscope", with a position with respect to the Moon identical to the position of the ascendant with respect to the Sun.[17]

---

[15] Britton-Walker 1999, p. 46.

[16] Diodorus, *Bibliotheca historica*, 2.47.6.

[17] For the 'Lot of Fortune', see in particolar Manilius, *Astronomica*, 3.160–202.

Chapter 13
Solar Eclipses, Lunar Eclipses and the Cycle of Saros

Eclipses occur when the Sun, Moon and Earth all come into alignment (Fig. 13.1). A solar eclipse occurs when the new Moon interposes itself between the Sun and the Earth (Figs. 13.2 and 13.3); a lunar eclipse occurs when the Earth comes between the Sun and the full Moon (Figs. 13.4 and 13.5).

If the Moon's orbit around the Earth was on the same plane as the Earth's orbit around the Sun, every new Moon would lie between the Sun and the Earth and we would have a solar eclipse, while at every full Moon the Earth would come between the Sun and the Moon, resulting in a lunar eclipse.

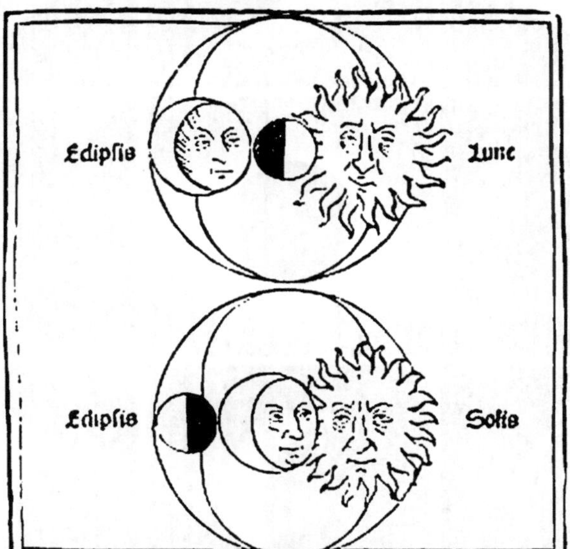

Fig. 13.1 Solar and Lunar eclipses—from Johannes de Sacrobosco (John of Holywood), *Sphaera Mundi*, Venice 1490. In Ptolemaic theory, the Earth lies at the centre of the universe: eclipses occur when the three heavenly bodies—the Earth, Moon and the Sun—come into line

L. Magini, *Stars, Myths and Rituals in Etruscan Rome*, Space and Society,
DOI 10.1007/978-3-319-07266-1_13

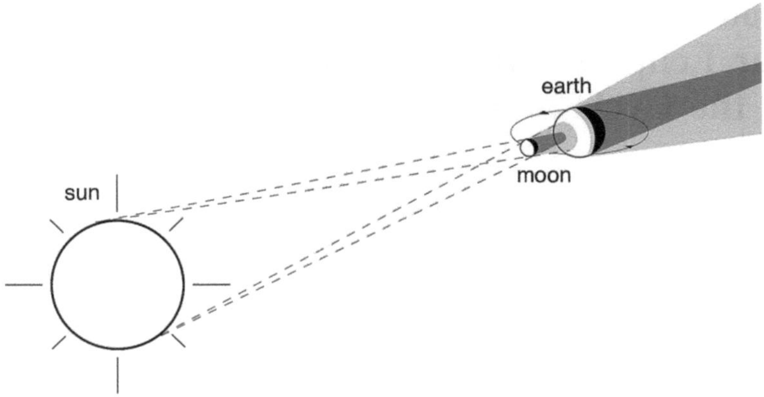

Fig. 13.2 Solar eclipse

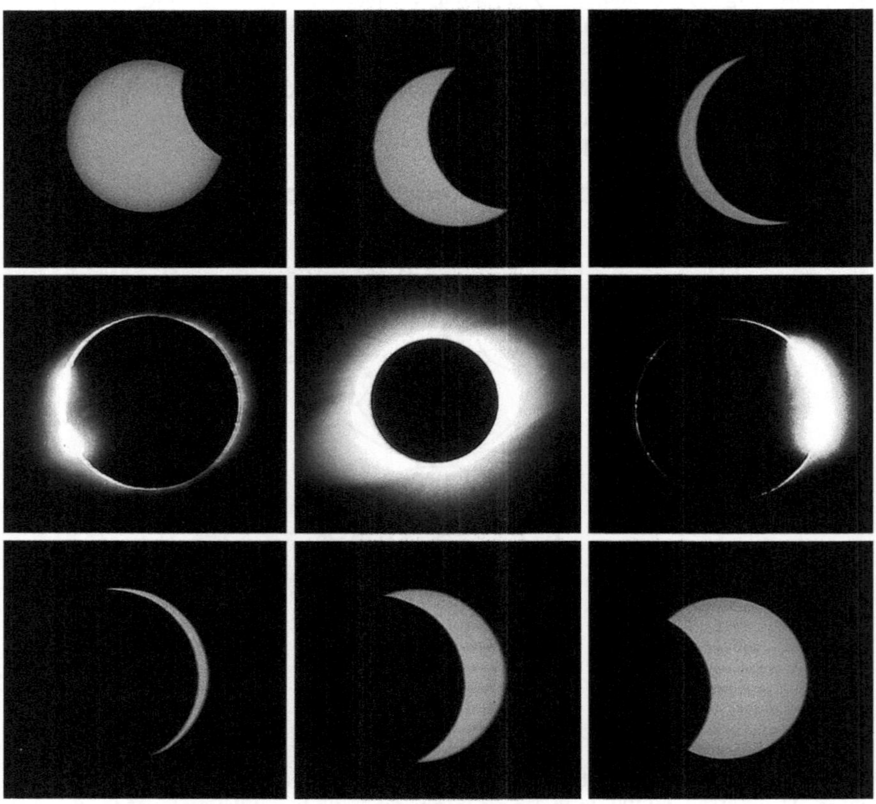

Fig. 13.3 Phases of the solar eclipse that occurred on 24 October 1995, photographed by G. Vanin

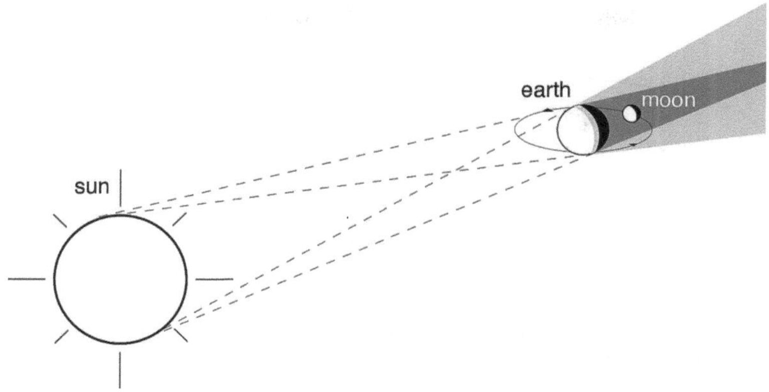

Fig. 13.4 Lunar eclipse

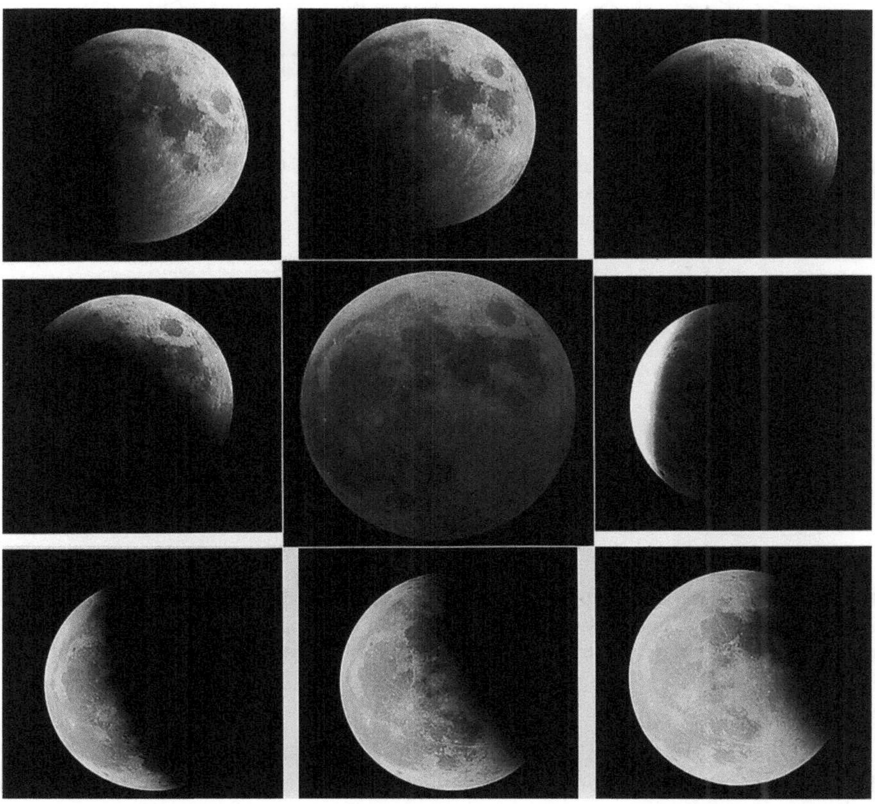

Fig. 13.5 Phases of the total lunar eclipse that occurred on 17 October 1986, photographed by C. Zanandrea

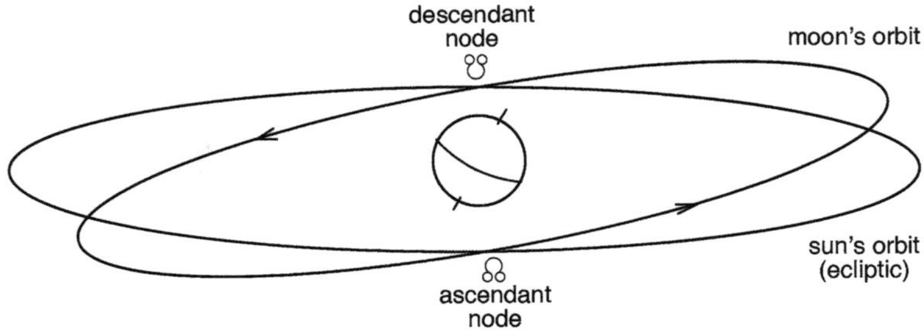

Fig. 13.6 The lunar nodes. The plane of the apparent orbit of the Moon is tilted by around 5° with respect to the plane of the apparent orbit of the Sun. The two lunar nodes (*ascendant and descendant*) are the points of intersection between the two orbits

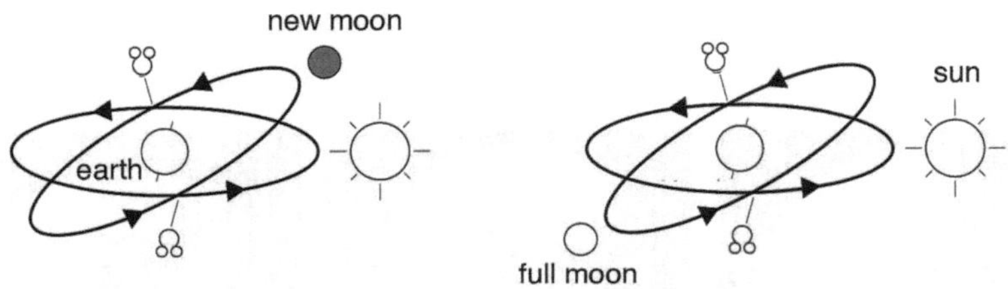

Fig. 13.7 When no eclipse takes place. The apparent orbits of the Sun and Moon and the two lunar nodes. *On the left*, the Moon is new; *on the right*, it is full; in both cases, there is no eclipse because the Sun and Moon are distant from the nodes

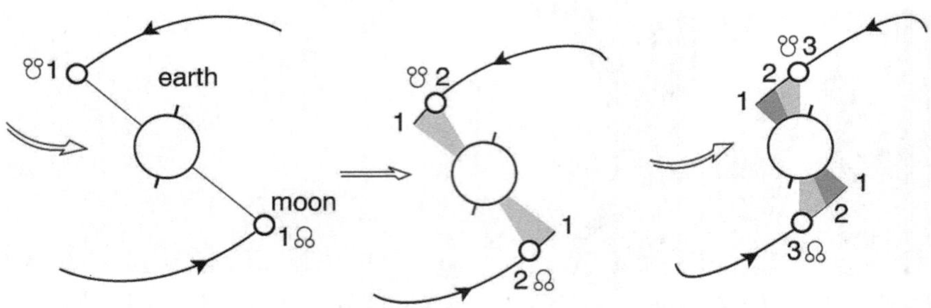

Fig. 13.8 The retrograde revolution of the lunar nodes along the ecliptic. The moon revolves around the Earth in an *anticlockwise direction*; the lunar nodes move in a *clockwise direction*. At each revolution the Moon passes the same node again before passing the same point in its own orbit

In actual fact, the Moon's orbit around the Earth is inclined with respect to the Earth's orbit around the Sun. It follows that in order for an alignment to trigger an eclipse, the full or new Moon must be near one of the *lunar nodes*, either *ascendant* or *descendant*, where the Moon's orbit intersects the Sun's orbit (Fig. 13.6). In such cases the Sun, viewed from the Earth, is either close to the same node where the new Moon is in transit, or is close to the opposite node where the full Moon is in transit. If this is not the case, no eclipse will take place (Fig. 13.7).

A total solar eclipse occurs when the Moon is new less than 11 days before or after the Sun passes through the given node; a partial eclipse occurs if the Moon is new less than 19 days before or after this transit. Lunar eclipses are total if the Moon becomes full less than 5 days before or after the Sun moves through the node; they are partial if the Moon becomes full less than 13 days before or after this passage.

The Sun takes 346.62 days—known as a *draconitic* or *eclipse year*—to return to the same lunar node. It takes half that time—173.31 days—to travel from one node to the next. Every year, eclipses are grouped into two *eclipse seasons* roughly six lunations apart. The Moon takes a *synodic month* of 29.5306 days to return to the same position with respect to the Sun—from new Moon to new Moon, or from full Moon to full Moon; it takes a *draconitic month* of 27.2122 days to return to the same node.

Eclipse years and draconitic months are shorter, respectively, than solar years and synodic months, because the lunar nodes, which spin in a retrograde—i.e. clockwise—direction along the ecliptic (Fig. 13.8), are, figuratively speaking, moving towards the Sun and the Moon, which, on the contrary, spin in a direct—i.e. anticlockwise—direction[1] (see Chap. 17).

The 6,585-day Saros cycle is the least common multiple of the synodic and draconitic months (see Table 13.1), that is to say, it is the shortest period of time between two moments when lunar and solar eclipses occur at practically the same frequency. By complete chance, it is also a multiple of the *sidereal month*, the *anomalistic month*—i.e. the interval between two successive passages of the Moon through perigee (see Chap. 19)—and the *eclipse year* (see again Table 13.1).

The Saros cycle—a phenomenon that was known in Mesopotamia, centuries before the start of the common era[2] (Fig. 13.9)—encompasses, on average, a total of 85 (total and partial) eclipses: 48 solar eclipses and 37 lunar eclipses.

Table 13.1 The Saros cycle

6,585.3238 days = 223 synodic months × 29.5306 days
6,585.3524 days = 242 draconitic months × 27.2122 days
6,584.5056 days = 241 sidereal months × 27.3216
6,585.5494 days = 239 anomalistic months × 27.5546 days
6,585.78 days = 19 eclipse years × 346.62 days

[1] The expressions "retrograde—i.e. clockwise—direction" and "direct—i.e. anticlockwise—direction" refer to an observer in the Earth's northern hemisphere looking towards the south.

[2] Schiaparelli (1997–1998, Vol. I, pp. 41–89) explains that the use of a lunar calendar—such as the Babylonian calendar—in which lunar eclipses can only occur mid-month makes it far easier to understand the Saros cycle.
Clearly enough, this explanation also applies to the Numan calendar, in which the underlying lunar foundation is disturbed only by intercalations. (see Chap. 15).

X 32 IV dir X	X 4 IV X	dir XI 1Ar IV X	XI 5 V dir XI	XI 11 V XI	dir XII 29 V XI
33 II 5 m VIII 34 II dir VIII 35 I VII 36 I VII	5 III 5 m dir IX 6 II VIII 7 II VIII 8 II VIdir VII	2 III 5 m IX 1Da III dir IX 2 II VIII 3 II VIII	6 III 5 m IX 1An III IX 2 III dir IX 3 II VIII	12 IV 5 m (dir) X 13 III IX 14 III IX 15 III dir IX	30 IV 5 m X 31 IV dir X 32 III IX 33 III IX
XII 5 m 37 VI dir XII 38 V XI 39 V XI	XII 5 m 9 VI XII 10 VI dir XII 11 V XI	4 I 5 m VIa XII 5 VI XII 1A VI dir XII	4 I 5 m VII 5 I VIa XII 6 VI XII	16 I 5 m VII 17 I VII 18 I VIa XII	34 II 5 m dir VIII 25 I
40 IV 5 m dir X 41 III IX 42 III IX 43 III dir IX	12 IV 5 m X 13 IV dir X 14 III IX 15 III IX	2 IV 5 m X 3 IV X 4 IV dir X 5 III IX	1Si V 5 m dir XI 2 IV X 3 IV X 4 IV dir X	19 V 5 m XI 20 V dir XI 21 IV X 22 IV X	
44 I 5 m VII 45 I VII XIIa 46 VI XII 1U VI	16 II 5 m dir VIII 17 I VII 18 I VII XIIa 19 VI	6 II 5 m VIII 7 II dir VIII 1Phi I VII 2 I dir VII	5 II 5 m VIII 6 II VIII 7 II dir VIII 8 I VII	23 II 5 m dir IX 24 II VIII 25 II VIII 26 II VIII	
XI 5 m 2 V dir XI 3 IV	XI 5 m 20 V XI 21 V	XII 5 m 3 V XI 4 V	XII 5 m 9 VI dir XII 10 V	XIIa 5 m 27 VI XII 28 VI	

Fig. 13.9 The so-called "Saros-Canon", a Babylonian tablet dating from the third century BCE, is a list of years and months listing eclipses of the Moon. The years are those of the successive reigns of Artaxerxes II (starting from the 32nd year), Umasu (U), Arses (Ar), Darius (Da), Alexander (A), Philip (Phi), Antigonus (An), and Seleucus (Si), from 373 to 277 BCE. The months are indicated as Roman numerals from I to XII; *dir* indicates the insertion of an intercalary month; 5 m the five-month interval between one series of eclipses and the next. Each column records 38 eclipses, corresponding to one cycle of Saros, composed of five series of 7, 8, 7, 8 and 8 eclipses (with the first series of eclipses subdivided into two portions of 3 and of 4 eclipses, respectively in the first and last row of boxes from the top). Within each series the interval between one eclipse and the next is 6 months (e.g. in the second square from the top of the first column: 33rd year of Artaxerxes, second and 8 month, and then in the 34th year, 2nd month, and so on)

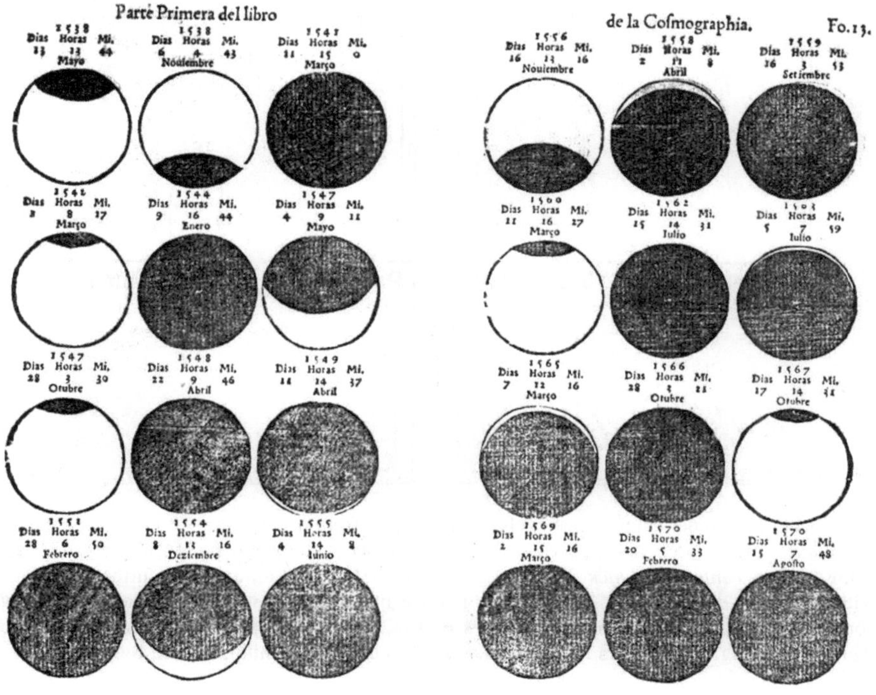

Fig. 13.10 Subsequent lunar eclipses visible from North Europe, predicted between the years 1538 and 1570—from Petrus Apianus (Peter Bienewitz), *Cosmographia*, Paris 1551. In the *second row, from left*, the eclipse on 1 March 1542 corresponds to the 12 March 1560 eclipse in the next Cycle of Saros

Solar eclipses are more numerous than lunar eclipses because two solar eclipses may potentially occur in a single eclipse season. Lunar eclipses, which may only occur once each time that the Sun passes through a node, are grouped into five series: three series lasting eight eclipses, and two lasting seven eclipses, in a pattern of 8, 7, 8, 8, 7. The last of these 38 eclipses is a repetition of the first: it marks the end of one Saros cycle and the beginning of the next.

Within an individual series, each eclipse occurs six lunations after the preceding eclipse; each series is separated by five lunations. Only the central eclipses in each series are total eclipses. The sequence of solar and lunar eclipses moves forward like toothed sprockets: one type of eclipse is total during the intervals between the other type of eclipse's series, which, vice versa, are either partial or fail to occur at all.[3]

At the end of each Saros cycle, the first eclipse is almost an exact repetition of the eclipse that occurred 223 synodic months and 242 draconitic months earlier (see Fig. 13.10). I say "almost exact" because the Saros cycle lasts a non-whole

[3] We shall see later on (Chap. 15) how to calculate the occurrence of eclipses within a Saros cycle.

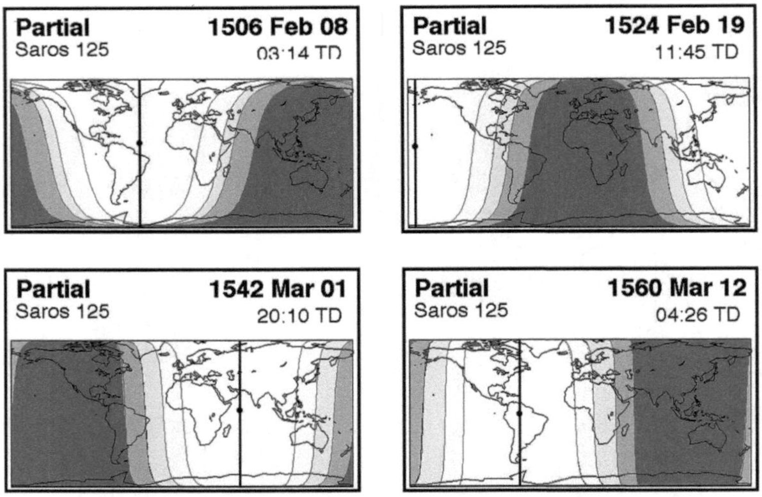

Fig. 13.11 Four lunar eclipses one cycle of Saros apart from one another—8 February 1506, 19 February 1524, 1 March 1542 and 12 March 1560—as shown today on the NASA web site. Each eclipse has its *centre* (*the black dot along the black vertical line*) around 120° further West than with preceding eclipse. The fourth eclipse is once more *centred* where the first was *centred*, completing the *exeligmos*. The two eclipses of 1542 and 1560 are among those predicted by Apianus; see Fig. 13.10 (from http://eclipse.gsfc.nasa.gov/eclipse.html, revised by the author)

number of days—6,585 and a third days, to be precise. In practice, this leads to a change in the area on Earth from which eclipses are visible between one Saros cycle and the next (see Fig. 13.11).

However, as Ptolemy, Geminus, Hipparchus, Aristarchus and, before them, astronomers in Babylon were well aware, adding three Saros cycles together forms a new 19,756 day-long cycle (close to a whole day),[4] known as an *Exeligmos*, or "that which has completely gone round" (see again Fig. 13.11).

[4] An *Exeligmos* lasts 669 synodic months, or 19,755.971 days; it also lasts 726 draconitic months, or 19,756.057 days.

Chapter 14
The Saros Cycle and the Feast Days of *Feriae Martis* and *Anna Perenna*

Six thousand, five hundred and eighty five days—one Saros Cycle!—elapse between 1 March in year one of the Numan cycle and 15 March in year nineteen (see Tables 4.1 and S. C.). The feast days celebrated on these two dates herald two conceptions and two solar eclipses.

The first of March in year one marks the anniversary of Romulus's conception and two simultaneous conjunctions (on Earth, between Mars and Rea Silvia, and in the skies, between the Sun and Moon). The latter conjunction resulted in the total eclipse of the Sun that marked the start of astronomical time and the Numan cycle (see Chap. 1).

The anniversary of another divine conception occurs one Saros cycle later, on 15 March in year nineteen:

> On the Ides is held the conception feast of Anna Perenna,[1]

writes Ovid, who knows a thing or two about the goddess:

> [...] but since erroneous rumours are rife as to who this goddess is,
> I am resolved to throw no cloak about her tale.

This is the poet's incipit for a sequence of stories that, as a whole, serve only as a framework, before going on to explain the reason for the vulgar songs sung by young women during the feast, when he comes to the crunch:

> Now it remains for me to explain why girls chant ribald songs;
> for they assemble and sing certain scurrilous verses.
> When Anna had been but lately made a goddess, the Marching God [i.e. Mars]
> came to her, and taking her aside spoke as follows:
> 'Thou art worshipped in my month, I have joined my season to thine:

---

[1] Ovid, *Fasti*, 3.523: *Idibus est Annae festum geniale Perennae.* The standard translation is by J.G. Frazer, who interprets the Latin term *geniale* as "jovial"—he is mistaken: the adjective *geniale* comes from the same root as the verb *gignere*, "to generate", and means "of generation, of conception". In *Ars Amatoria*, 1.125, Ovid writes *genialis praeda* in reference to the Sabine virgins who have been chosen to conceive children for the Romans.

L. Magini, *Stars, Myths and Rituals in Etruscan Rome*, Space and Society,
DOI 10.1007/978-3-319-07266-1_14

I have great hope in the service that thou canst render me.
An armed god myself, I have fallen in love with the armed goddess Minerva;
I burn and for a long time have nursed this wound.
She and I are deities alike in our pursuits; contrive to unite us.
That office well befits thee, kind old dame.'
So he spoke. She duped the god by a false promise,
and kept him dangling on in foolish hope by dubious delays.
When he often pressed her, 'I have done thy bidding,' said she,
'she is conquered and has yielded at last to thine entreaties.'
The lover believed her and made ready the bridal chamber. Thither
they escorted Anna, like a bride, with a veil upon her face.
When he would have kissed her, Mars suddenly perceived Anna;
now shame, now anger moved the god befooled.
The new goddess laughed at dear Minerva's lover.
Never did anything please Venus more than that.
So old jokes are cracked and ribald songs are sung,
and people love to remember how Anna choused the great god.[2]

As in one of the paintings that brought such renown to the Italian Renaissance, poet Ovid offers an allegorical description of a heavenly body coming between two others. In this particular case, the story illustrates how Anna Perenna, interposing herself between the ardent god Mars and the coveted Minerva, repeats on Earth the acts of the celestial bodies on high: Mars embodies the male element and represents the Sun, the infinite and everlasting fiery source of seed, while Minerva embodies the female element, in representation of the Moon, the no less infinite and everlasting aqueous reservoir of the ovum.[3]

Countless attempts by the female element (Anna or Minerva) to escape a tryst with a male element (represented by Mars) are correlated by countless conjunctions or oppositions (new Moons or full Moons) in which the two brightest bodies in the sky come into close proximity without actually generating an eclipse. When,

---

[2] Ovid, *Fasti*, 3.543-4 and 675-96: *Quae tamen haec dea sit, quoniam rumoribus errat, / fabula proposito nulla tegenda meo […] / Nunc mihi cur cantent superest oscena puellae / dicere; nam coeunt certaque probra canunt. / Nuper erat dea facta; venit Gradivus ad Annam / et cum seducta talia verba facit: / 'Mense meo coleris, iunxi mea tempora tecum; / pendet ab officio spes mihi magna tuo. / Armifer armiferae correptus amore Minervae / uror et hoc longo tempore volnus alo. / Effice, di studio similes coeamus in unum. / Conveniunt partes haec tibi, comis anus.' / Dixerat; illa deum promisso ludit inani / et stultam dubia spem trahit usque mora. / Saepius instanti: 'Mandata peregimus', inquit, / 'evicta est; precibus vix dedit illa manus.' / Credit amans thalamosque parat. Deducitur illuc / Anna tegens voltus, ut nova nupta, suos. / Oscula sumpturus subito Mars aspicit Annam: / nunc pudor elusum, nunc subit ira deum. / Ludis amatorem; cara es, nova diva, Minervae / nec res hac Veneri gratior ulla fuit. / Inde ioci veteres obscenaque dicta canuntur / et iuvat hanc magno verba dedisse deo.*

[3] Plutarch, *de Facie*, 934C and 944E, explains that the Moon, "if she is eclipsed when dawn is already near, she takes on a bluish or azure hue, from which especially it is that the poets… give her the epithet 'bright-eyed' [or 'blue-eyed']. […] for it must be out of love for the Sun that the Moon herself goes her rounds and gets into conjunction with him in her yearning [to receive] from him what is most fructifying". "Bright-eyed" or "blue-eyed"—better, "Owl-eyed" (*Glaukopis*)—is an attribute of Athena, the Greek corollary to the Etruscan/Roman goddess Minerva.

at last, as they travel along their respective orbits, the Sun and Moon both approach the lunar nodes and an eclipse becomes possible (see Chaps. 13 and 15), this is the moment when (in mythical terms) Anna Perenna comes between Mars and Minerva, corresponding (in astronomical terms) to the "full Earth" interposing itself between the Sun and the Moon, and blotting out the full Moon during lunar eclipse.

This explains why, when Ovid writes of the newly-elevated goddess,

> Some think that this goddess is the Moon, because the Moon
> [fills up the measure of the year by her months,[4]

he uses the term *annus* in its acceptation of "temporal circuit", its oldest meaning. Macrobius bears this out:

> In reality, as the lunar year is in fact the month, for in a little less than a month the Moon travels around the entire Zodiac, so the solar year should be calculated as the number of days that it takes to return to the same sign from which the Sun departed. This is known as the course of the year—the grand year; the Moon's year, on the contrary, is the abbreviated year. Ateius Capito is another who believes that the noun *annus* derives from 'temporal circuit'.[5]

In actual fact, on 15 March in year nineteen of the Numan cycle, 6,585 days later, the Moon concludes the temporal circuit defined by the Saros cycle. On that day, however, the Moon is new—and not full, as the Ides on the calendar would seem to indicate—so the eclipse is once more a solar eclipse that falls on the day of the conception of Anna Perenna, repeating the 1 March solar eclipse in year one, and marking the start of the next Saros cycle.

The countless jugs of wine downed and the (same) number of years that participants at the feast wish one another to live mark a rare, exceptional and extraordinary event that takes place once in a very great number of years (and glasses of wine)….

At the precise moment when, in the heavens, one astral cycle comes to an end and the next is about to begin, down on Earth another cycle of life comes to an end: the cycle of adolescence, which is replaced by the cycle of youth.

This is the moment when young women go through their initiation rites (see Chap. 11) and prepare for marriage in the near future, on *Veneralia*, 1 April (see Chap. 10). Males complete their cycle of adolescence and enter youth at the end of *tirocinium*, which is followed by enrollment in the Army (for those who are eligible), the donning of the virile toga, and presentation to the people on 17 March, on the feasts of *Liberalia* (see Chaps. 7 and 11), while their ever more

---

[4] *Fasti*, 3.523-696. Quoted here is line 657: *Sunt quibus haec Luna est, quia mensibus impleat annum.*

[5] Macrobius, *Saturnalia*, 1.14.4-5: *Nam, sicut lunaris annus mensis est, quia luna paulo minus quam mensem in zodiaci circumitione consumit, ita solis annus hoc dierum numero colligendus est quem peragit dum ad signum se denuo vertit ex quo digressus est; unde annus vertens vocatur, et habetur magnus, cum lunae annus brevis putetur […] Hinc et Ateius Capito annum a circuitu temporis putat dictum.*

precocious female contemporaries do so 2 days earlier, on 15 March, the feast of *Anna Perenna*.

This is the moment when virgins are inspected to ensure that they are *viripotens*, "of marriageable age" and *integra*, "intact, inviolate"; precisely because these inspections are yet to be completed and impending marriages are yet to be celebrated, the virgins in question are not allowed to copulate.[6]

This is when virgins visit the Lavinium temple, bringing goblets dedicated to *Ana* and other objects such as statues and votive objects to propitiate their fertility as newlyweds. This is when virgins "meet to sing vulgar songs"—a typical pre-wedding event, a kind of hen-party—and the blood of their final menstruation before consummation of matrimony is poured on the sacred orchard of the goddess, along the Via Flaminia, just outside the modern day Porta del Popolo in Rome.

According to Martial,

> […] where virginal blood assists
> Anna Perenna's woods, so heavy with fruit,[7]

in which—if we so desire—we may perceive an arch reference to the female body and the "fruits" enjoyed by the man who penetrates her "thicket".

The feast of *Anna Perenna* is the moment when the cycle of life that runs from birth to the age of fertility cedes to a no less life-filled cycle, moving forwards from the start of fertility to the birth of the next generation of humankind.

---

[6] See Torelli 1984, pp. 57–71 and 105–15; a different view.

[7] Martial, *Epigrams*, 4.64.16-7: *et quod virgineo cruore gaudet / Annae pomiferum nemus Perennae.*

Chapter 15
The Frequency of Solar and Lunar Eclipses and the Festivities of the Numan Year

If we set the Saros and Numan cycles in motion concurrently, we find that:

- the total eclipse of the Sun that took place on Romulus's conception (1 March in year one of the Numan cycle) kicks off astronomical and calendar time in both cycles;
- the total eclipse of the Sun 6,585 days later that took place on Anna Perenna's conception (15 March in year nineteen of the same Numan cycle) marks the end of a first Saros cycle and the start of the next.[1]

It is an interesting exercise to calculate the solar and lunar eclipses within the Saros cycle, match them to the Numan cycle, and compare these dates with feast days in the Numan calendar. As we have seen, it is straightforward enough to make such calculations approximately by positing that the Earth's revolution around the Sun and the Moon's revolution around the Earth are uniform in motion and follow circular orbits, and considering the apparent dimension of the Sun and Moon to be their average size. Though things in the natural world are different—as ancient sky gazers became only too aware at a very early stage—the margin of error generated by this kind of simplification is not nearly significant enough to prevent construction of a theoretical calendar that could be used to predict eclipses.

The key calculation for solar eclipses is the moment that the Sun passes through one of the lunar nodes, and the number of days between this moment and the next new Moon. This calculation is most straightforward if we follow *the simplest hypothesis: on the first day of the cycle the Sun passes through a node, and the Moon is in the new Moon phase.*

[1] The following calculations are restricted to the duration of the Saros cycle, in part as a result of doubts regarding about intercalations during the final 8 years of the Numan cycle (see Chap. 4); the only requisite for these calculations to remain valid is for year eighteen not to be an intercalated year (see Table 4.1).

L. Magini, *Stars, Myths and Rituals in Etruscan Rome*, Space and Society,
DOI 10.1007/978-3-319-07266-1_15

We therefore start with a total solar eclipse on the first day of the cycle. Having passed through a node on the first day of the cycle, the Sun moves through the next node 173.31 days later; the Moon—which is new on the first day of the cycle—completes six lunations 177.18 (= 6 × 29.53) days later. The Moon is new 3.87 (= 177.18 − 173.31) days after the Sun passed through the node. In consequence, a second solar eclipse occurs during the sixth new Moon after the first solar eclipse, 177 days later, on 26 August in year one. This too is a total eclipse, because it occurs within the 11 day interval of the Sun's passage through the node.

The Sun next passes through a node 173.31 days later, 346.62 days after the start of the cycle. The Moon takes 354.36 days to complete 12 lunations. The Moon is new 7.74 (= 354.36 − 346.62) days after the Sun passes through the node, which means that the twelfth new Moon coincides with a third solar eclipse on the 354th day, 27 February (the last but 1 day of the year one). Once again, this is a total eclipse, because the Sun passed through the node fewer than 11 days earlier.

Another 173.31 days later (519.93 days after the start of the cycle), the Sun passes through a node once more. The Moon completes 18 lunations after 531.55 days, and is new 11.62 days after the Sun's passage through the node. The eighteenth new Moon heralds a fourth solar eclipse, this time not total, because it is more than 11 days after the Sun's passage through the node.

From now on, the Sun is too far from the node for the alignment between the Sun, Moon and Earth to occur: the shadow of the new Moon no longer strikes the surface of the Earth, and the sequence of eclipses is temporarily suspended.

However, five lunations later—on the occasion of the twenty-third new Moon, 679.19 days after the beginning of the cycle—the Sun approaches a node for the fourth time (693.24 days from the start). The Moon is new 14.05 days before the Sun moves through the node. As this is within the threshold of 19 days, we have a partial solar eclipse that starts off a new series of eclipses. And so the sequence continues.

There are, of course, occasions when it is necessary to calculate the number of days between the Sun's passage through a node and the two nearest new Moons: one prior to and one subsequent to the Sun's passage. This allows us to ascertain whether or not two consecutive new Moons fall within a single eclipse season. If we take the previous example, the Sun passes through a node for the third time after 519.93 days; the seventeenth lunation is complete after 502.02 days, and the eighteenth lunation after 531.54 days. Consequently, the seventeenth new Moon occurs 17.91 (= 519.93 − 502.02) days earlier, and the eighteenth new Moon takes place 11.62 (= 531.55 − 519.93) days after the Sun passes through a node, giving us two (extremely) partial solar eclipses, one on the seventeenth and the other on the eighteenth new Moon. And so the sequence continues.

The same process applies to lunar eclipses, with the sole difference that the calculation concerns the number of days between the Sun's passage through one of the nodes and the full Moon, i.e. the half-point in a lunation.

After calculating eclipse dates and the Saros cycle day number when these eclipses occur, we must check the day, month and year in the Numan cycle to which they correspond. This may be achieved from the years of the Numan cycle (see Table 4.1) and months of the Numan year (see S. C.).

Tables 15.1 and 15.2 compare the characteristics of the Saros cycle with the Numan cycle. As we saw in Chap. 4, the Numan cycle consists of 12-lunation-long years in odd years of the cycle, alternating with longer, intercalated even years. From one year to the next, without intercalation eclipses would occur on the same day (or at most a day earlier); with intercalation eclipses slip backwards by the number of intercalated days (22 or 23 days, or at most a day earlier).

On occasion, an eclipse may reoccur on the same day 10 years later. For example, the 22nd lunar eclipse on 11 August in year eleven reflects the first lunar eclipse on 11 August in year one. It is self-evident why this occurs: the first eclipse coincides with the 129th full Moon, and the second eclipse coincides with the 5th full Moon of the cycle. Through intercalation, 124 (= 129 − 5) lunations or 3,662 days elapse over the 10 years separating the two eclipses (see Chap. 4). I say "on occasion" because the five-lunation interval between one series of eclipses and the next (see Chap. 13) breaks up the regularity of eclipses falling on the same day in later years. For instance, though the lunar eclipse on 11 August in year one is followed by a lunar eclipse on 11 August in year eleven, the 13 February eclipse in year one is not mirrored by a 13 February eclipse in year eleven because a new series commenced one lunation earlier, on 12 January.

Note

The following two Tables list solar and lunar eclipses in the Saros cycle for the first 19 years of a Numan cycle: Table 15.1 is a list of solar eclipses (standard typeface); Table 15.2 is a list of lunar eclipses (italics).
Total eclipses are in bold; partial eclipses are in standard typeface:
Column A measures the Sun's passages through one of the lunar nodes and the lunations for a solar eclipse (lunation midpoints for lunar eclipses);
Column B shows the day that the Sun passes through one of the nodes, the day of lunation (or lunation midpoint), and the number of days between the two;
Column C presents the eclipse day (the lunation day, rounded up or down);
Column D lists the day, month and year of the Numan cycle in which the eclipse occurs;
Column E shows Numan calendar feasts (*Feriale antiquissimum*[2] feasts are in bold) that fall exactly on the same day as an eclipse (in which case the day is not indicated) or close to that day. For simplicity's sake, months are abbreviated in accordance with our modern-day calendar (e.g. March = 3).
A thicker horizontal line separates the series of solar eclipses; a double line is used to separate the lunar eclipse series.[3]

[2] See Chap. 2.

[3] The list of total solar eclipses in Table 15.1 (and subsequent tables) are those that arise from calculations performed as described above, circumscribed to the eclipse season as defined in Chap. 13; in consequence, they are "theoretical eclipses". Total lunar eclipse dates are also theoretical. It should be noted, however, that these calculations incorporate the corrections proposed by Pannekoek 1989, p. 59, based on the actual variation of the Moon's angle and the Sun's velocity, and their distribution within each of the Saros series.

Table 15.1 Solar eclipses in the Saros cycle and the initial 19 years of the Numan cycle

A	B	C	D	E
No. of the Sun's passages through a node/No. of lunations	Day of the Sun's passage through a node—Day of lunation = Number of days between the Sun's passage through a node and the lunation	Eclipse day (rounded up or down to the nearest day)	Numan cycle day, month and year	Numan calendar feast day
0/0	**1 − 1 = 0**	**1**	**1 Mar. 01**	**FERIAE MARTIS**
1/6	**173.31 − 177.18 = −3.87**	**177**	**26 Aug. 01**	**25.8. OPIC.-27.8. VOLT.**
2/12	**346.62 − 354.37 = −7.75**	**354**	**27 Feb. 01**	**EQUIRRIA**
3/17	519.93 − 502.02 = +17.91	502	27 Jul. 02	25.7. FURRINALIA
3/18	519.93 − 531.55 = −11.62	532	26 Aug. 02	25.8. OPIC.-27.8. VOLT.
4/23	693.24 − 679.20 = +14.04	679	26 Jan. 02	27.1. Castoribus
4/24	693.24 − 708.73 = −15.49	709	4 intercal. 02	* * *
5/29	**866.55 − 856.39 = +10.16**	**856**	**4 Jul. 03**	**5.7. POPLIFUGIA**
6/35	**1,039.86 − 1,033.57 = +6.29**	**1,034**	**4 Jan. 03**	**3.-5.1. Laralia**
7/41	**1,213.17 − 1,210.75 = +2.42**	**1,211**	**4 Jul. 04**	**5.7. POPLIFUGIA**
8/47	**1,386.48 − 1,387.94 = −1.46**	**1,388**	**3 Jan. 04**	**3.-5.1. Laralia**
9/53	**1,559.79 − 1,565.09 = −5.30**	**1,565**	**9 Jun. 05**	**VESTALIA**
10/59	**1,733.10 − 1,742.31 = −9.21**	**1,742**	**8 Dec. 05**	**Gaiae**
11/64	1,906.41 − 1,889.96 = +16.45	1,890	10 May 06	9.-11.-13.5 LEMURIA
11/65	1,906.41 − 1,919.49 = −13.08	1,919	8 Jun. 06	9.6. VESTALIA
12/70	2,079.72 − 2,067.14 = +12.58	2,067	7 Nov. 06	4.-7.11. Ludi Plebei
12/71	2,079.72 − 2,096.67 = −16.95	2,097	8 Dec. 06	Gaiae
13/76	**2,253.03 − 2,244.33 = +8.70**	**2,244**	**16 Apr. 07**	**15.4. FORDICIDIA**
14/82	**2,426.34 − 2,421.51 = +4.83**	**2,421**	**15 Oct. 07**	**October Equus/IDUS IOVI**
15/88	**2,599.65 − 2,598.69 = +0.96**	**2,599**	**16 Apr. 08**	**15.4. FORDICIDIA**
16/94	**2,772.96 − 2,775.88 = −2.92**	**2,776**	**15 Oct. 08**	**October Equus/IDUS IOVI**
17/100	**2,946.27 − 2,953.06 = −6.79**	**2,953**	**23 Mar. 09**	**TUBILUSTRIUM**
18/105	3,119.58 − 3,100.71 = +18.87	3,101	20 Aug. 09	19.8. VINALIA
18/106	3,119.58 − 3,130.24 = −10.66	3,130	20 Sep. 09	* * *
19/111	3,292.89 − 3,277.90 = +14.95	3,278	21 Feb. 09	FERALIA
19/112	3,292.89 − 3,307.43 = −14.54	3,307	22 Mar. 10	23.3. TUBILUSTRIUM
20/117	3,466.20 − 3,455.08 = +11.12	3,455	19 Aug. 10	VINALIA
21/123	**3,639.51 − 3,632.26 = +7.25**	**3,632**	**20 Feb. 10**	**21.2. FERALIA**
22/129	**3,812.82 − 3,809.45 = +3.37**	**3,809**	**27 Jul. 11**	**25.7. FURRINALIA**
23/135	**3,986.13 − 3,986.63 = −0.50**	**3,987**	**27 Jan. 11**	**Castoribus**
24/141	**4,159.44 − 4,163.81 = −4.37**	**4,164**	**27 Jul. 12**	**25.7. FURRINALIA**
25/147	**4,332.75 − 4,341.00 = −8.25**	**4,341**	**26 Jan. 12**	**27.1. Castoribus**
26/152	4,506.06 − 4,488.65 = +17.41	4,489	3 Jun. 13	1.6. Iunoni Monetae
26/153	4,506.06 − 4,518.18 = −12.12	4,518	3 Jul. 13	5.7. POPLIFUGIA

(continued)

Table 15.1 (continued)

A	B	C	D	E
No. of the Sun's passages through a node/No. of lunations	Day of the Sun's passage through a node—Day of lunation = Number of days between the Sun's passage through a node and the lunation	Eclipse day (rounded up or down to the nearest day)	Numan cycle day, month and year	Numan calendar feast day
27/158	4,679.37 − 4,665.83 = +13.54	4,666	2 Dec. 13	1.12. Neptunus
27/159	4,679.37 − 4,695.37 = −16.00	4,695	2 Jan. 13	3.-5.1. Laralia
28/164	**4,852.68 − 4,843.02 = +9.66**	**4,843**	**2 Jun. 14**	**1.6. Iunoni Monetae**
29/170	**5,025.99 − 5,020.20 = +5.79**	**5,020**	**1 Dec. 14**	**Neptuno**
30/176	**5,199.30 − 5,197.39 = +1.91**	**5,197**	**10 May 15**	**9.-11.-13.5. LEMURIA**
31/182	**5,372.61 − 5,374.57 = −1.96**	**5,374**	**7 Nov. 15**	**4.-17.11. Ludi Plebei**
32/188	**5,545.92 − 5,551.75 = −5.83**	**5,552**	**10 May 16**	**9.-11.-13.5. LEMURIA**
33/194	**5,719.23 − 5,728.94 = −9.71**	**5,729**	**7 Nov. 16**	**4.-17.11. Ludi Plebei**
34/199	5,892.54 − 5,876.59 = +15.95	5,876	16 Mar. 17	Annae P./IDUS IOVI
34/200	5,892.54 − 5,906.12 = −13.58	5,906	15 Apr. 17	FORDICIDIA
35/205	6,065.85 − 6,053.77 = +12.12	6,054	14 Sep. 17	13.9. IOVI EPULUM
35/206	6,065.85 − 6,083.30 = −17.45	6,083	14 Oct. 17	15.10. Octob. Eq./IDUS IOVI
36/211	**6,239.16 − 6,230.96 = +8.20**	**6,231**	**16 Mar. 18**	**15.3. Annae P./IDUS IOVI**
37/217	**6,412.47 − 6,408.14 = +4.33**	**6,408**	**13 Sep. 18**	**IOVI EPULUM**
38/223	**6,585.78 − 6,585.32 = +0.46**	**6,585**	**15 Mar. 19**	**Annae Peren./IDUS IOVI**

Table 15.2 Lunar eclipses in the Saros cycle and the initial 19 years of the Numan cycle

A	B	C	D	E
No. of the Sun's passages through a node/No. of lunation midpoints	Day of the Sun's passage through a node—Day of lunation midpoint = Number of days between the Sun's passage through a node and the lunation midpoint	Eclipse day (rounded up or down to the nearest day)	Numan cycle day, month and year	Numan calendar feast day
1/5.5	173.31 − 162.42 = +0.89	162	11 Aug. 01	10.8. Opis et Cereris
2/11.5	346.62 − 339.60 = +7.02	340	13 Feb. 01	Parentalia/IDUS IOVI
3/17.5	519.93 − 516.79 = +3.16	517	11 Aug. 02	10.8. Opis et Cereris
4/23.5	**693.24 − 693.97 = −0.73**	**694**	**12 Feb. 02**	**13.2. Parentalia/IDUS IOVI**

(continued)

Table 15.2 (continued)

A	B	C	D	E
No. of the Sun's passages through a node/No. of lunation midpoints	Day of the Sun's passage through a node—Day of lunation midpoint = Number of days between the Sun's passage through a node and the lunation midpoint	Eclipse day (rounded up or down to the nearest day)	Numan cycle day, month and year	Numan calendar feast day
5/29.5	**866.55 − 871.15 = −4.60**	**871**	**19 Jul. 03**	**LUCARIA**
6/35.5	1,039.86 − 1,048.34 = −8.48	1,048	18 Jan. 03	* * *
7/41.5	1,213.17 − 1,225.52 = −12.35	1,225	18 Jul. 04	19.7. LUCARIA
8/46.5	1,386.48 − 1,373.17 = +13.31	1,373	17 Dec. 04	SATURNALIA
9/52.5	1,559.79 − 1,550.36 = +9.43	1,550	25 May 05	Fortunae Publicae
10/58.5	**1,733.10 − 1,727.54 = +5.56**	**1,728**	**22 Nov. 05**	* * *
11/64.5	**1,906.41 − 1,904.72 = +1.69**	**1,905**	**25 May 06**	**Fortunae Publicae**
12/70.5	**2,079.72 − 2,081.91 = −2.19**	**2,082**	**22 Nov. 06**	
13/76.5	**2,253.03 − 2,259.09 = −6.06**	**2,259**	**2 May 07**	**1.5. Laribus/Bonae Deae**
14/82.5	2,426.34 − 2,436.27 = −9.93	2,436	30 Oct. 07	* * *
15/88.5	2,599.65 − 2,613.46 = −13.81	2,613	1 May 08	Laribus/Bonae Deae
16/93.5	2,772.96 − 2,761.11 = +11.85	2,761	29 Sep. 08	1.10. Tigillo sororio
17/99.5	2,946.27 − 2,938.29 = +7.98	2,938	8 Mar. 09	7.3. Vediovi
18/105.5	**3,119.58 − 3,115.48 = +4.10**	**3,115**	**5 Sep. 09**	**Ludi Magni**
19/111.5	**3,292.89 − 3,292.66 = +0.23**	**3,293**	**8 Mar. 10**	**7.3. Vediovi**
20/117.5	**3,466.20 − 3,469.85 = −3.65**	**3,470**	**5 Sep. 10**	**Ludi Magni**
21/123.5	3,639.51 − 3,647.03 = −7.52	3,647	12 interc. 10	* * *
22/129.5	3,812.82 − 3,824.21 = −11.39	3,824	11 Aug. 11	10.8. Opis et Cereris
23/134.5	3,986.13 − 3,971.87 = +14.26	3,972	12 Jan. 11	11.1. Iuturn./CARMENTALIA
24/140.5	4,159.44 − 4,149.05 = +10.39	4,149	12 Jul. 12	13.7. Ludi Apollinares
25/146.5	**4,332.75 − 4,326.23 = +6.52**	**4,326**	**11 Jan. 12**	**11.1. Iuturn./CARMENT.**
26/152.5	**4,506.06 − 4,503.42 = +2.64**	**4,503**	**17 Jun. 13**	**18.6. Annae sacrum**
27/158.5	**4,679.37 − 4,680.60 = −1.23**	**4,681**	**17 Dec. 13**	**SATURNALIA**

(continued)

Table 15.2 (continued)

A	B	C	D	E
No. of the Sun's passages through a node/No. of lunation midpoints	Day of the Sun's passage through a node—Day of lunation midpoint = Number of days between the Sun's passage through a node and the lunation midpoint	Eclipse day (rounded up or down to the nearest day)	Numan cycle day, month and year	Numan calendar feast day
28/164.5	**4,852.68 − 4,857.78 = −5.10**	**4,858**	**17 Jun. 14**	**18.6. Annae sacrum**
29/170.5	5,025.99 − 5,034.97 = −8.98	5,035	16 Dec. 14	17.12. SATURNALIA
30/176.5	5,199.30 − 5,212.15 = −12.85	5,212	25 May 15	Fortunae Publicae
31/181.5	5,372.61 − 5,359.80 = +12.81	5,360	24 Oct. 15	* * *
32/187.5	5,545.92 − 5,536.99 = +8.93	5,537	24 Apr. 16	23.4. VINALIA/25.4. ROBIG.
33/193.5	**5,719.23 − 5,714.17 = +5.06**	**5,714**	**23 Oct. 16**	* * *
34/199.5	**5,892.54 − 5,891.35 = +1.19**	**5,891**	**31 Mar. 17**	**Lunae**
35/205.5	**6,065.85 − 6,068.54 = −2.69**	**6,068**	**28 Sep. 17**	**1.10. Tigillo sororio**
36/211.5	6,239.16 − 6,245.72 = −6.56	6,246	31 Mar. 18	Lunae
37/217.5	6,412.47 − 6,422.91 = −10.44	6,423	28 Sep. 18	1.10. Tigillo sororio

Another thing that Table 15.1 reveals is the correspondence between eclipse days and *Feriale antiquissimum* feast days in the Numan calendar:

- 10 solar eclipses fall exactly on a feast day: on 27 February 01, 9 June 05, 15 October 07, 15 October 08, 23 March 09, 21 February 09, 19 August 10, 15 April 17, 13 September 18, and 15 March 19 (the last eclipse in the first Saros cycle, and the first in the second);
- 17 solar eclipses fall at a one-day interval—either before or after—a feast day: on 26 August 01, 26 August 02, 4 July 03, 4 July 04, 10 May 06, 8 June 06, 16 April 07, 16 April 08, 20 August 09, 22 March 10, 20 February 10, 10 May 15, 10 May 16, 16 March 17, 14 September 17, 14 October 17, and 16 March 18;
- 4 solar eclipses fall at a 2 day interval from a feast day: on 27 July 02, 27 July 11, 27 July 12, and 3 July 13.

In total, 31 (= 10 + 17 + 4) eclipses out of 49 fall on or around *Feriale antiquissimum* feast days.

Incidentally, two issues should be borne in mind if we are to correctly interpret what these results mean; unfortunately, there is a great deal more to be learned about these two issues, not least the way the transition was made in ancient times from theory to practical application.

The first issue is Etrusco-Roman civilization's predilection for odd numbers, and its corresponding dislike of even numbers (see Chap. 2). Given this knowledge, we may surmise that even if the calculations reveal that an eclipse fell on an even-numbered day, the corresponding festivity would be inserted into the calendar on the previous or subsequent odd-numbered day.

The second issue regards Etruscan and Roman differences in what is deemed to be the start of the day. Virgil's commentator informs us: "The day begins at midday for the Etruscans, and at midnight for the Romans."[4] From this, it is clear that what in the modern age is the morning was, for the Etruscans, the last half of the day (and for the Romans the first half of the day); and what in the modern age is the afternoon was, for the Etruscans, the first half of the day (and for the Romans the last half of the day). In other words, for example, our morning of day 7 was already the first half of day 7 for the Romans, whereas for the Etruscans it was still the second half of day 6.

The first of these issues potentially leads to confusion over the roles of even-numbered and odd-numbered days; the second issue may lead to confusion over the role of the first and second half of the day.

If we return to the calculations in Table 15.1, the first 10 eclipses "take up"—if such a term may be permitted – 10 days in the calendar; the second 17 eclipses, which fall either one day before or one day after a feast day, take up 51 (= 17 eclipses × 3 days) days; the final 4 eclipses, which falls either 2 days before or 2 days after the feast day, take up 20 (= 4 × 5) days. All in all, the 31 (= 10 + 17 + 4) eclipses take up 81 (= 10 + 51 + 20) days out of the 355 days in the year.

If we compare the 81 days taken up by eclipses during a 355 day year with the 6,585 days in the Saros cycle, we obtain the following ratio: 81 ÷ 355 = 1,502 ÷ 6,585 = 0.228. This means that around 1,502 days are taken up by eclipses in one Saros cycle, making the theoretical probability that an eclipse falls on any given day 22.8 %. And yet the number of eclipses that fall on *Feriale antiquissimum* days are 31 ÷ 49, corresponding to 63.3 %. In other words, during the Saros cycle that corresponds to the first 19 years of the Numan cycle, the frequency with which solar eclipses occur on *Feriale antiquissimum* feast days is almost three times the theoretical frequency (63.3 % as against 22.8 %).

[4] Servius, *ad Aen.*, 5.738: *hic autem dies […] inchoat […] secundum Etruscos […] a sexta hora diei, secundum Romanos a media nocte.*

There is another, parallel line of reasoning that allows us to obtain the same result: if the theoretical probability that an eclipse falls on any feast day is equal to 22.8 %, it requires more than 4 eclipses to reach 100 % certainty. But if the number of eclipses in one Saros cycle totals 49 in all, around 12 of them (= 49 ÷ 4) should theoretically fall on a feast day; and yet 31 eclipses fall on feast days, more than three times the theoretical probability.

If we now consider just the total eclipses in Table 15.1, we find that 18 out of 28 eclipses fall on feast days, corresponding to a percentage of 64.3 %:

- 7 eclipses fall precisely on a feast day: on 27 February 01, 9 June 05, 15 October 07, 15 October 08, 23 March 09, 13 September 18, and 15 March 19;
- 9 eclipses fall at a one-day interval—either before or after—from a feast day: on 26 August 01, 4 July 03, 4 July 04, 16 April 07, 16 April 08, 20 February 10, 10 May 15, 10 May 16 and 16 March 18;
- 2 eclipses fall at a two-day interval from a feast day: on 27 July 11, 27 July 12.

All in all, 44 (= 7 + 9 eclipses × 3 days + 2 eclipses × 5 days) days out of the 355 day year are taken up with eclipses, corresponding to a percentage of 12.4 %. Eclipses fall on feast days more than five times more frequently than theoretical probability: 64.3 compared with 12.4.[5]

This is a clear demonstration that the nameless creator of the Numan calendar and cycle possessed considerable astronomical and mathematical knowledge.

For a counterproof that these calculations are correct, it is sufficient to compare the dates of the theoretical eclipses calculated above with the historical eclipses recorded at and around the turn of the 21st century. According to the ephemeris,[6] if we designate the first total solar eclipse in an imaginary modern-day Numan cycle as the eclipse recorded on 18 March 1988, and then calculate the total and partial solar and lunar eclipses that occurred over the following 6,585 days, up to 29 March 2006, we obtain the following results (see Table 15.3).

The theoretical series we have composed and the historical series concord perfectly, both in terms of dates and eclipse types. The few exceptions that occur crop up because the historical eclipse that starts off our series, which occurred on 18 March 1988, took place when the Sun and Moon were around 4.5° from the node, rather that at the node itself.

[5] These calculations have been revised and completed on the basis of observations by Lewis Licht and John T. Ramsey in their review of my book *Astronomy and Calendar in Ancient Rome—The Eclipse Festivals*, in *Archaeoastronomy*, Vol. 18 (2004), pp. 124–126. The book was also reviewed by Germaine Aujac in *Cairn—Revue de philologie de littérature et d'histoire anciennes*, 2001, LXXV, pp. 170–172, and by Matthew F. Dowd in *Centaurus*, 2002, Vol. 44, pp. 143–144.

[6] *The Complete Ephemerides 1920–2020*, edited by F. Santoni, Aureas Editions, Paris 1995.

Returning to the relationship between eclipses and Roman feast days (see Tables 15.1 and 15.2), we may note that a large number of the 84 solar and lunar eclipses in the Saros cycle that begins on 1 March in year one of the Numan cycle and concludes on 15 March of year nineteen, occur (some of them multiple times) on the same Numan year feast day. It follows that *certain days in the Numan year become days for celebrating the anniversary of an eclipse* even during years when no eclipse takes place (see Table 15.4).

Table 15.3 Theoretical and historical eclipses

Theoretical Eclipses	Historical Eclipses			Theoretical Eclipses	Historical Eclipses		
Day of theoretical eclipse	Day of historical eclipse	Type of historical eclipse		Day of theoretical eclipse	Day of historical eclipse	Type of historical eclipse	
1	**1**	○	**Tot.**	3,307			
162	*162*	●	*Par.*	3,455	3,455	○	Par.
177	**177**	○	**Tot.**	***3,470***	***3,469***	●	***Tot.***
340	*339*	●	***Tot.***	**3,632**	**3,632**	○	**Tot.**
354	354	○	Par.	*3,647*	*3,647*	●	*Par.*
502					*3,795*	●	Par.
517	***517***	●	***Tot.***	**3,809**	**3,809**	○	**Tot.**
532	531	○	Par.	*3,824*	*3,824*	●	*Par.*
679	**679**	○	**Tot.**	*3,972*	*3,971*	●	*Par.*
694	***693***	●	***Tot.***	**3,987**	**3,987**	○	**Tot.**
709				*4,149*	*4,149*	●	*Par.*
856	**856**	○	**Tot.**	**4,164**	**4,163**	○	**Tot.**
871	*871*	●	*Par.*	***4,326***	***4,326***	●	***Tot.***
1,034	**1,033**	○	**Tot.**	**4,341**	4,341	○	Par.
1,048	*1,048*	●	*Par.*	4,489	4,488	○	Par.
	1,196	●	*Par.*	***4,503***	***4,503***	●	***Tot.***
1,211	**1,210**	○	**Tot.**	4,518	4,518	○	Par.
1,225	*1,225*	●	*Par.*	4,666	4,665	○	Par.
1,373	*1,373*	●	*Par.*	***4,681***	***4,680***	●	***Tot.***
1,388	**1,387**	○	**Tot.**	4,695			
1,550	*1,550*	●	*Par.*	**4,843**	**4,843**	○	**Tot.**
1,565	**1,565**	○	**Tot.**	***4,858***	*4,857*	●	*Par.*
1,728	***1,727***	●	***Tot.***	**5,020**	**5,019**	○	**Tot.**

(continued)

Table 15.3 (continued)

Theoretical Eclipses	Historical Eclipses			Theoretical Eclipses	Historical Eclipses		
Day of theoretical eclipse	Day of historical eclipse	Type of historical eclipse		Day of theoretical eclipse	Day of historical eclipse	Type of historical eclipse	
1,742	1,742	○	Par.	*5,035*	*5,035*	●	*Par.*
1,890	1,890	○	Par.		*5,182*	●	*Par.*
1,905	***1,904***	●	***Tot.***	**5,197**	**5,197**		**Tot.**
1,919				*5,212*	*5,211*	●	*Par.*
2,067	2,066	○	Par.	*5,360*	*5,360*	●	*Par.*
2,082	***2,082***	●	***Tot.***	**5,374**	**5,374**	○	**Tot.**
2,097				5,537	***5,537***	●	***Tot.***
2,244	**2,244**	○	**Tot.**	**5,552**	**5,552**	○	**Tot.**
2,259	*2,259*	●	*Par.*	***5,714***	***5,714***	●	***Tot.***
2,421	**2,421**	○	**Tot.**	**5,729**	**5,728**	○	**Tot.**
2,436	*2,436*	●	*Par.*	5,876	5,876	○	Par.
	2,584	●	*Par.*	***5,891***	***5,891***	●	***Tot.***
2,599	**2,598**	○	**Tot.**	5,906			
2,613				6,054	6,054	○	Par.
2,761	*2,760*	●	*Par.*	***6,068***	***6,068***	●	***Tot.***
2,776	**2,776**	○	**Tot.**	6,083			
2,938	***2,938***	●	***Tot.***	**6,231**	**6,230**	○	**Tot.**
2,953	2,952	○	Par.	*6,246*	*6,246*	●	*Par.*
3,101				**6,408**	**6,408**	○	**Tot.**
3,115	***3,115***	●	***Tot.***	*6,423*	*6,422*	●	*Par.*
3,130	3,130	○	Par.		6,570	●	*Par.*
3,278	**3,278**	○	**Tot.**	**6,585**	**6,585**	○	**Tot.**
3,293	*3,293*	●	*Par.*				

Solar eclipse (○), *lunar eclipse* (●). The theoretical eclipses column replicates Column C (Eclipse days) in Tables 15.1 and 15.2

A few things are worth noting. First, the type of feast days associated with eclipses: *Laralia*, *Lemuria*, *Feralia*, *Parentalia*, *Laribus* are all celebrations of ancestors, the deceased and the dead, which in some way are associated with the afterlife. This confirms that eclipses represent a moment when the exceptional configuration of the celestial bodies in the sky favours communication between the macro- and microcosm in both directions, upwards and downwards.

The second fact of note regards the names of feast days and/or the divinities celebrated. The above-mentioned feasts—along with others such as *Volturnus*, *Furrina*, *Neptunus*, *Fortuna*, *Iuturna*, *Carmenta* and *Saturnus*—are of Etruscan origin. Could this be a clue to the origin of the "barbarian superior to Pythagoras" who schooled King Numa?

The third fact regards female initiation ceremonies. The *Tigillum sororium* feast on 1 October—the date of the mythical duel between the Horatii and the Curiatii—is associated with the rites of passage (initiation rites for male and female adolescents) and marks the beginning of the process. The feast day of 15 March, *Anna Perenna*, in which "virginal blood is poured on the goddess's sacred wood", heralds the end of the process, at least as far as women are concerned (see Chap. 14). The former of these feast days coincides with the 16th, 35th and 37th lunar eclipses; the latter coincides with the 47th and 49th solar eclipses (see Tables 15.1 and 15.2). Half a month after the conclusion of female initiation rites is the *Veneralia*, celebrated on 1 April, when Roman women celebrate marriage; in the most auspicious of years, the planet Venus/Hesperus is in lower conjunction with the Sun (see Chap. 10). As Manilius writes,

> Who after this can doubt that a link exists between heaven and man?[7]

And yet Manilius penned this lines several centuries after somebody had taught Numa how it all worked.

The fourth thing worth noting is how the Saros cycle concludes. The 48th total eclipse of the Sun—i.e. the last but one of the cycle (see Table 15.1)—just so happens to fall on the most important feast day of the whole year. The *Epulum Iovi Iunoni Minervae*, on 13 September, recalls the foundation, the *dies natalis*, of the temple dedicated to the Capitoline Triad. The high point of the feast is the moment when earthly people dare to touch the statue of the god with a human hand as they paint him with minium, symbolically representing the return to full power of the Sun as it leaves its temporary state of eclipse.

Before we conclude, there is one more (seemingly) spurious comment to add: if we had followed the inverse procedure to the one described above, we would have obtained the same result. In effect, if we had hypothesized that the feast days in the Numan calendar mark eclipses, we could have taken a Numan year with the *Feriale antiquissimum* festivities, inserted it into the appropriate cycle with the appropriate intercalations, totted up the interval of time between a certain number of these festivities, and obtained the interval of time between the various eclipses calculated within a Saros cycle. These intervals of time—characteristic of the

---

[7] Manilius, *Astronomica*, 2.105: *quis dubitet post haec hominem coniungere caelo?*

Table 15.4 The Saros cycle in the first 19 years of the Numan cycle

Cycle years	Mar.	Apr.	May	Jun.	Jul.	Aug.	Sep.	Oct.	Nov.	Dec.	Jan.	Feb.
1	○					●○						●○
2					○	●○					○	●○
3					○●						○●	
4					○●					●	○	
5			●	○					●	○		
6			○●	○					○●	○		
7		○	●					○●				
8		○	●				●	○				
9	●○						●○					○
10	●○					○	●					○●
11					○	●					●○	
12					●						●○	
13				○●	○					○●	○	
14				○●						○●		
15			○●					●	○			
16		●	○					●	○			
17	○●	○					○●	○				
18	○●						○●					
19	○											

Solar eclipse (○), lunar eclipse (●)

cycle of eclipses—would have been "conclusive" to recognizing the existence of such a cycle. As, indeed, they were not so many years ago, in interpreting the *Table of Eclipses in* the Mayan "Dresden Codex".[8]

One final, fundamental thing must be noted: lunar eclipses are visible from the entire half of the globe which, at the time of the eclipse, is in darkness. The same, however, does not apply to solar eclipses, which are only visible along a thin, long strip around the half of the Earth in daylight at the time of the eclipse. In consequence, out of all the solar and lunar eclipses calculated for the first 19 years of the Numan cycle, only around half of the lunar and an extremely small number—perhaps even none—of the solar eclipses would have been visible from Rome.

[8] See Aveni, 2000, pp. 139–144.

This confirms what we said at the beginning: for at least some of its feast days, the *Numan calendar is a theoretical calendar that could be used to predict eclipses*: it "could be used" even if the predicted eclipse would not necessarily be visible from Rome itself, particularly if the eclipse was a solar eclipse.

Chapter 16
Roman and Mesopotamian Eclipses: The Feast Days of the *Regifugium* and *Vestalia*

Since the most ancient of times, eclipses—solar eclipses in particular—have been viewed as harbingers of misfortune. It was for this reason that in Babylonian times enormous efforts were made to predict eclipses and avert their most ill-starred effects on the King, who was the earthly representative of the god who is represented in the sky by the Sun. The existence of a Mesopotamian ritual for warding off the evil consequences of eclipses (documented particularly under the reigns of Asarhaddon from 680 to 669 BCE, and Assurbanipal from 668 to 627 BCE) and the legendary tale, traditionally dated to 509 BCE, of the Kings being chased out of Rome reveal a similar conceptual approach by two civilizations which, as history would have it, never came into contact.

In the Babylonian *sar puhi* ritual—the term literally means "substitute King"[1]—during the immediate lead-up to an eclipse announced by obscure presages and predicted by high-priest astronomers as fatal for the King, the King was replaced by a *saklu*,

> a term which, in general, designates a common, ingenuous man who may even be somewhat simple […] without social importance, whose destiny is of little interest to anyone.

The *saklu*, who potentially may have heard tell of a prophecy that he would assume regal powers, dons the garb and badges of royalty, enters the palace, sits on the throne, and occupies the royal bed with a virgin married specially for the occasion. His reign lasts "three months and ten days". In the meantime, while continuing his de facto reign, the real King resides in a sacred enclosure—known as a *qersu*, or "block, reed hedge"—while he waits for the danger represented by the eclipse to pass. After 100 days, the *saklu* and his unlucky companion are put to death, their clothes and baubles burnt, and

---

[1] See Bottéro 1991, pp. 145–165; this documentation and translated sources are his.

L. Magini, *Stars, Myths and Rituals in Etruscan Rome*, Space and Society,
DOI 10.1007/978-3-319-07266-1_16

> the country and the King both go through a general process of purification […] To finish this off, a deep 'cleanse' is required of the palace, after its apparent 'contamination' by the presence of the now defunct stand-in. This operation was undertaken in six phases (carried out, perhaps, over the same number of days […]), each of which regarded one of the building's key areas: four were for the main entrances, one for the 'court room' which, in all likelihood, was the throne room, and the other was for the *Divan*, where during the day the king granted audiences, received his subjects, made and made public his rulings […] Only then would the King return to his palace and once more take up his role and public life. The death of his substitute marked the end of any danger to his person and, in consequence, to his people and his country.

The Roman version runs thus:

- in the *Regifugium*, on 24 February, the king makes a sacrifice at the *Comitium* and then flees,[2] to be substituted by the *interrex*, the "inter-king", who takes his place during the final five days of February;
- there was a solar eclipse on 27 February in year one of the Numan cycle, 12 lunations after the eclipse of 1 March (see Table 15.1);
- the last Etruscan king of Rome, Tarquin the Superb (535–509 BCE), had evil omens including

> a horrid sight! from between the altars a snake
> came forth and snatched the sacrificial meat from the dead fires,[3]

and in a dream the King comes upon a ram which charges and knocks him over. Then, Tarquin,

> prostrate on the ground, wounded gravely,
> raises his eyes to the heavens to witness an event immense
> and full of wonder: the Sun's flaming disc
> spews forth rays and melts away to its right,
> reversing its path across the heavens.[4]

The Roman version also has its "simpleton", who is

> the prudent Brutus [who] feigned to be a fool,
> in order that from thy snares, Tarquin the Proud, dread king, he might be safe;

and it is Brutus, the one whom the prophecy predicts:

> 'He who shall first have kissed his mother will be victorious',

---

[2] Plutarch, *Quaestiones Romanae*, 63.

[3] The *Regifugium* is narrated by Ovid in *Fasti*, 2.685-852. Quoted here are lines 711-2: *[…] nefas visu, mediis altaribus anguis / exit et extinctis ignibus exta rapit.*

[4] Cicero, *de divinatione*, 1.44, citing Accio's *Brutus*: *exim prostratum terra, graviter saucium, / resupinum in caelo contueri maxumum ac / mirificum facinus: dextrorsum orbem flammeum / radiatum solis liquier cursu novo.*

as he is the only one to realize that the "mother" in question is Mother Earth. It is Brutus once more who avenged the outrage perpetrated on Lucretia,

> as the Sun was making ready to hide his face,[5]

and who expelled Tarquin from the throne, "substituting him" once and for all with the revolutionary *sar puhi* that marked the birth of the Republic, and which is recalled by the *Regifugium* feast day.

All of this occurs during the five days that follow on from the *Terminalia* on 23 February, as we saw in Chaps. 4 and 7:

> The *Terminalia* [are so called] because this day is set as the last day of the year; for the twelfth month was February, and when the extra month is inserted the last 5 days are taken off the twelfth month.[6]

If, then, starting from the *Terminalia* "the last 5 days are taken off" from February, what remains are 31 days in March, 29 in April, 31 in May and 9 in June to make one hundred days, or the "three months and ten days" during which the King—in Babylonia—remained isolated in his "reed hedge", for the length of time that he was imperilled by the eclipse.

In Rome, around 9 June, a purification ceremony was held for

> this little spot, which now supports the Hall of Vesta,
> [that] was then the great palace of unshorn Numa.

The ceremony lasted a whole novena: it began on 7 June, when the house of the Goddess was thrown open next to the *Regia* (Fig. 16.1); it reached its height on 9 June[7] and concluded on 15 June with two days of initial and final purification, both of which share the same nundinal letter: C. The final day is, of course,

> the day on which thou, O Tiber, dost send the filth of Vesta's temple
> down the Etruscan water to the sea.[8]

---

[5] Ovid, *Fasti*, 2.717-8, 713-4, 786: *Brutus erat stulti sapiens imitator, ut esset / tutus ab insidiis, dire Superbe, tuis; […] 'Matri / qui dederit princeps oscula victor erit.'; […] condere iam voltus sole parante suos.*
The Latin *stultus* specifically indicates a "fool or imbecile"; then again, the very name *Brutus* means "dumb (like an animal)".

[6] Varro, *de lingua latina*, 6.13; see n. 7.1.

[7] 9 June is the date of the 12th total solar eclipse, on the 1,565th day of the Saros cycle; 27 February is the date of the third, on the 354th day (see Table 15.1). Therefore, a total of 1,211 (= 1,565 – 354) days elapse between these two eclipses, or if you prefer, 41 lunations. It transpires that one of the two intervals of time after which an eclipse is almost always repeated is 41 lunations (the other is 47 lunations).

[8] Ovid, *Fasti*, 6.263-4 and 713-4: *hic locus exiguus, qui sustinet Atria Vestae, / tunc erat intonsi regia magna Numae; […] haec est illa dies, qua tu purgamina Vestae, / Thybri, per Etruscas in mare mittis aquas.*

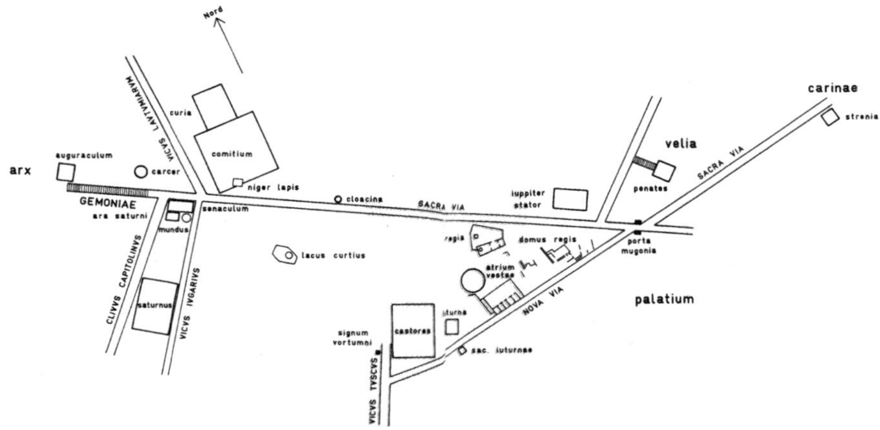

Fig. 16.1 A schematic reconstruction of Central Ancient Rome. The *House of Vesta* is near the *Royal Palace* (see this chapter.). The temple of Saturn is near the base of the Capitol (see Chap. 23), not far from the *Mundus* which coincides with the *Umbilicus Urbis* (see Chap. 23). The temple of *Ianus Geminus* (see Chap. 23) should be located midway between the *niger lapis* and the temple of *Cloacina*. (from Coarelli 1983)

Varro has the following to say about 15 June:

> The day which is called 'When the dung has been carried out, Right,' is named from this, that on this day the dung is swept out of the Temple of Vesta and is carried away along the Capitoline Incline to a certain spot.[9]

One may legitimately wonder why such a ritual was institutionalized in Rome where its underlying reason simply did not pertain: there is no lack of firewood in Rome or in the Latium region, and there is no need to go looking for other raw materials to put on the fire, though that could not be said about Ancient or modern Mesopotamia. In a modern-day account of Iraq we read the following:

> Girls help their mothers or older sisters gather fuel (brushwood or dried animal dung), as well as keep the house supplied with water, and cook.[10]

It is almost as if we obtain a glimpse, living and breathing, of the last heirs to the Vestals, as we follow a trajectory the origins of which is lost in the long night of history… though *not* Roman history.

The same might be said of the relationship between the Babylonian *sar puhi* and Rome's *Regifugium*. How can the latter possibly be independent of the former? And if it is not independent, how ever did it make the journey from Babylon to Rome?

---

[9] Varro, *de lingua latina*, 6.32: *Dies qui vocatur 'Quando Stercus Delatum Fas', ab eo appellatus, quod eo die ex aede Vestae stercus everritur et per Capitolinum Clivum in locum defertur certum.*

[10] Fiorina 1985, pp. 72–74.

Chapter 17
The Revolution of the Line of Nodes and the Feast Days of *October Equus*, *Fordicidia* and *Parilia*

The *lunar nodes* are the points where the Moon's apparent orbit, which is inclined by 5.1° with respect to the ecliptic, intersects the Sun's apparent orbit. The nodes are referred to as either the *ascendant node* or the *descendant node*, depending upon whether, during its transit, the Moon moves from a negative latitude to a positive latitude, or vice versa (see Chap. 13 and Fig. 13.6). The line that joins these two nodes is known as the *line of nodes*.

According to astrological tradition, the two nodes are known as the *Head* and *Tail of the Dragon*, because the ancients imagined that a giant dragon lay along the ecliptic, its head on the ascendant node and its tail on the descendant node (Figs. 17.1 and 17.2), poised to cause much-feared eclipses by devouring either the Sun or the Moon.

As we have seen (see Chap. 13), the nodes rotate along the ecliptic in a retrograde direction, "heading towards" the Sun and the Moon, which rotate in direct motion. It therefore follows that the Sun and the Moon pass back through a given node before they pass back through the same point on their respective orbits (see again Fig. 13.8): the *draconitic year*—also known as the *eclipse year*—is 346.62 days long, while the *draconitic month* lasts 27.2122 days.

The lunar nodes—and, in consequence, the *line of nodes*—complete their retrograde revolution with respect to the fixed stars in around 6,793 days, at a rate of around −0.0530 degrees per day.[1]

The 6,793th day in the Numan cycle corresponds to 14 October in the nineteenth year (see Table 4.1 and S. C.). On this day, the *Dragon's Head* and *Tail* return to the position they occupied on the first day of the cycle. The following day—15 October—in Ancient Rome was *October Equus*, a celebration at which a *horse's head* and *tail* were chopped off.[2]

[1] $0.0530 = 360 \div 6.793$.

[2] Polybius 12.4b; Plutarch, *Quaestiones Romanae*, 97; Festus, p. 190-1L1; Paulus Diaconus, p. 246L1.

L. Magini, *Stars, Myths and Rituals in Etruscan Rome*, Space and Society,
DOI 10.1007/978-3-319-07266-1_17

Fig. 17.1 The Head and Tail of the Dragon in a "modern" representation—from Albumasar, *de magnis coniunctionibus*, Augusta Vindelicorum, 1489

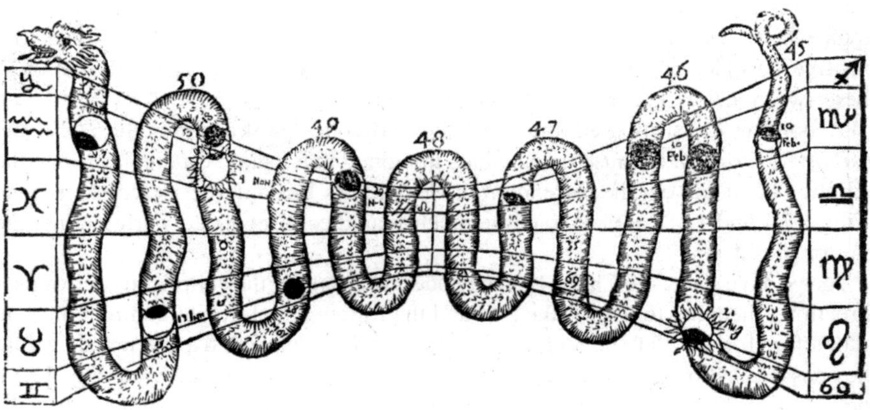

Fig. 17.2 The Moon's Nodes, i.e. the Head and Tail of the Dragon—from Athanasius Kircher, *Ars Magna Lucis et Umbrae*, Amsterdam 1671. This xylography explains how to calculate and predict eclipses: the lunar nodes—the Head and Tail of the Dragon—rotate, and eclipses occur alternately in the signs of Aquarius, Taurus, Capricorn, etc., up to Leo and Scorpio

Six lunations earlier or later[3] than the *October Equus*, on 15 April, is the feast day of *Fordicidia*. Half an eclipse year—in other words, 173 days—before the *October Equus*, on 21 April,[4] is when the feast of *Parilia* occurs.

---

[3] Counting inclusively, we have 179 days before and 178 days after 15 October: 179 days = 15 days in April + 149 in May, June, July, August and September + 15 in October; 178 days = 17 days in October + 146 in November, December, January, February and March + 15 in April.

[4] 173 days = 9 days in April + 149 in May, June, July, August and September + 15 in October.

These three celebrations share common elements in their rituals and astronomy:

- Rites: the blood from the tail of the horse sacrificed on 15 October, mixed with ashes from calves taken from their mothers' wombs on 15 April and empty, dried fava bean husks, is used for the suffumigation prepared by the Vestals for 21 April. Ovid writes:

> Ye people, go fetch materials for fumigation from the Virgin's altar.
> Vesta will give them; by Vesta's gift ye shall be pure.
> The materials for fumigation will be the blood of a horse and the ashes of a calf;
> the third thing will be the empty stalks of hard beans.[5]

- Astronomy: if one node coincides with the *October Equus*, the other coincides with the *Fordicidia*; if an eclipse occurs for the *October Equus*, it is likely that the previous or next eclipse will (have) fall(en) on the *Fordicidia*; if the Sun transits through a node at *Parilia*, it will transit through the other node on the *October Equus*.

The 15 April celebration is a feast of spring, to propitiate a more plentiful harvest and larger flocks and herds:

> […] the year proved more fruitful,
> and Earth and cattle yielded their increase.[6]

The 15 October celebration is an autumn feast,

> for the harvest received.[7]

The celebration on 21 April coincided with the traditional date of Rome's foundation. Logically, the node through which the Sun passed at that moment—in an ideal, as yet still to be identified year—should be the ascendant node, the *Head of the Dragon*, which "like a beneficent star […] signifies dominion, fortune and riches". By the following 15 October, the Sun would be transiting the descendant node, the *Tail of the Dragon*, which, "signifying degeneration, misfortune and poverty", and "cold and corruptive",[8] is, according to Pliny, similar to how the horse's blood "corrodes the ulcers and attacks the sides".[9]

With the right intercalation, three solar eclipses take place in concomitance with the *Fordicidia*, two of which are total (in year seven and year eight of the Numan cycle), plus one partial solar eclipse in year seventeen. The two total eclipses take place at one node, while the partial eclipse takes place at the other.

---

[5] Ovid, *Fasti*, 4.731-4: *I, pete virginea, populus, suffimen ab ara: / Vesta dabit, Vestae munere purus eris. / Sanguis equi suffimen erit vitulique favilla, / tertia res durae culmen inane fabae.*

[6] Ovid, *Fasti*, 4.671-2: *[…] fecundior annus / provenit, et fructum terra pecusque ferunt.*

[7] Paulus Diaconus, p. 246L1: *ob frugum eventum.*

[8] For the Head of the Dragon, please see: Alchabitius, *Libellus isagogigus* 1512, fo. 45r; for the Tail of the Dragon, see: *Ex libris Mysteriorum Apomasaris*, in CCAG XII, p. 102. Both of these texts may be found in Bezza, 1995, vol. 1, p. 412.

[9] Pliny, *Naturalis Historia*, 28.147: *erodit, emarginat ulcera.*

A further three solar eclipses take place in concomitance with the *October Equus*, in the same years, and with the same characteristics (see Table 15.1).

This information is insufficient to establish whether—in an as yet to be identified year, as noted earlier—the ascendant node through which the Sun passes at the *Fordicidia,* and therefore at the *Parilia* too, is that of the total eclipse of years seven and eight, or the partial eclipse of year seventeen. If we prefer the *hypothesis* of the pair of total eclipses, we find that the total eclipse of the Sun on 1 March in year one takes place with the Sun and Moon both near the descendant node (see Table 15.4).

Chapter 18
The Ascendant Node at the First Point of Aries and the Festival of *Summanus*

The Earth's annual revolution around the Sun takes place along the ecliptic plane. The Earth's daily rotation around its own axis takes place on the plane of the celestial equator, which is inclined 23.5° with respect to the ecliptic. In consequence, the ecliptic intersects the celestial equator at two points: the *First Point of Aries* (also known as the Gamma Point), and the *First Point of Libra*. Viewed from the Earth, the Sun passes through the First Point of Aries on 21 March, at the spring equinox, and through the First Point of Libra on 23 September, at the autumn equinox (see Fig. 18.1).

The retrograde revolution of the lunar nodes along the ecliptic means that the ascendant and descendant nodes alternately transit through the First Point of Aries roughly every 3,396 days.[1]

When the ascendant node transits through the First Point of Aries, during one half lunation the Moon's declination—that is to say, its height in the sky—diminishes dramatically, by more than 57°[2] (see Fig. 18.2).

In such cases, at the beginning of the lunation the initial crescent Moon rises and sets further north than the Sun does at the summer solstice; two weeks later, the full Moon rises and sets further south than the Sun at its winter solstice. The points on the horizon where the Moon rises and sets are known as the *upper* and *lower major standstills* (see Fig. 18.3).

When, on the contrary, the descendant node transits through the First Point of Aries, during one half lunation the Moon's declination diminishes by a less dramatic 37°[3] (see again Fig. 18.2). In such cases, at the beginning of the lunation the

---

[1] 3,396 = 6,793 ÷ 2.

[2] The Earth's axis is inclined 23.5° with respect to the ecliptic; the plane on which the Moon revolves around the Earth is inclined by 5.1° with respect to the ecliptic. When, as we find here, these two inclinations are added together, over the course of two weeks the Moon's declination varies from +28.6° (= 23.5° + 5.1°) to –28.6°, i.e. 57.2°.

[3] In such cases, the two inclinations (see preceding footnote) are subtracted; over the course of two weeks the Moon's declination varies from +18.4° (= 23.5° – 5.1°) to –18.4°, i.e. 36.8°.

L. Magini, *Stars, Myths and Rituals in Etruscan Rome*, Space and Society,
DOI 10.1007/978-3-319-07266-1_18

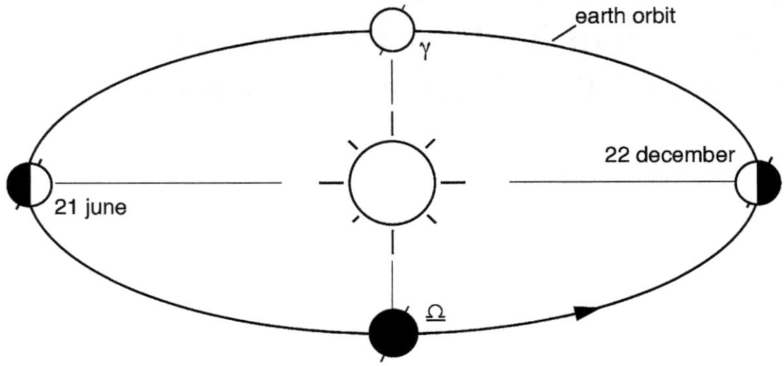

Fig. 18.1 The Earth's motion around the Sun. Viewed from the Earth, the Sun moves through the *First Point of Aries* on 21 March, the Spring Equinox, and through the *Point of Libra* on 23 September, the Autumn Equinox

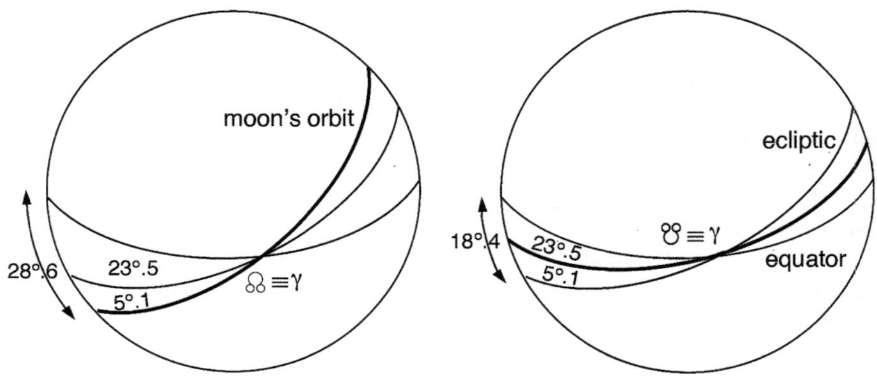

Fig. 18.2 The lunar nodes and their passage from the Points of Aries and Libra. When the ascendant node moves through the *First Point of Aries*, the Moon's declination diminishes by more than 57° during one half lunation; when the descendant node transits through the *First Point of Aries*, the Moon's declination diminishes by almost 37° during one half lunation. (from Romano 1985)

initial crescent Moon rises closer to the east and sets closer to the west than the Sun at its summer solstice; two weeks later, the full Moon rises closer to the east and sets closer to the west than the Sun at its winter solstice. The points on the horizon where the Moon rises and sets are known as the *upper* and *lower minor standstills* (see again Fig. 18.3).[4]

[4] See Romano 1994, pp. 163-6.

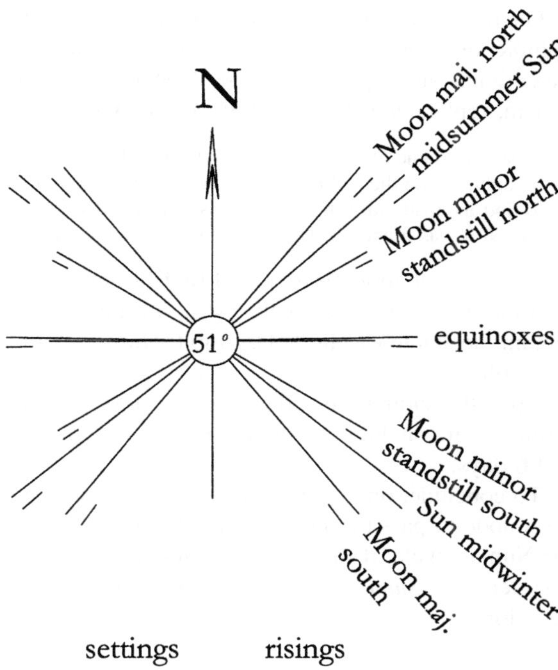

Fig. 18.3 The Moon's standstills. Major standstills and minor standstills (at a 51° North latitude, such as London or Paris). (from North 2008)

Let's suppose—as we have just seen—that on 1 March of year one in the Numan cycle, the Sun is close to the descendant node and is in eclipse (see Chap. 17 and Table 15.4), the new Moon will also be close to the descendant node. On that day, the node itself is close to the First Point of Aries, so the half lunation that commences sees the Moon's declination diminish by roughly 37°: the crescent Moon rises at the upper minor standstill and, fifteen days later, the full Moon rises at the lower minor standstill.

The ascendant node will consequently transit through the First Point of Aries 3,396 days later, on 20 June in year ten (see Tables 4.1 and S. C.), on the feast of *Summanus* and close to the summer solstice. The Moon is, once again, new[5]: the first crescent rises at the upper major standstill and, fifteen days later, the full Moon rises at the lower major standstill. Viewers from Earth witness an extraordinary and worrying spectacle as the Moon "falls" by over 57°.

Conversely, if we presume that on 1 March of year one in the Numan cycle, the Sun is close to the ascendant node, the two celestial events described above would take place on the same dates, but in an inverted manner: initially, on 1 March of year one, the more "dramatic" event; finally, on 20 June in year ten, the less dramatic.

[5] $3{,}396.019 = 115 \times 29.5306$.

On the latter day, the statue of the deity who falls from the heights of the Temple of Jupiter Capitoline, his head ending up in the river, re-evokes this exceptional and worrying celestial phenomenon and, using the power of symbolism, replicates it in an earthly setting. Cicero recalls the myth in these words:

> When the statue of Summanus, which was still made of terracotta at that time, was struck by lightning on the pediment of the temple of Jupiter Optimus Maximus, no one was able to find its head, until the soothsayers said it had been knocked off into the Tiber, where indeed it was found, in the exact spot the soothsayers predicted.[6]

This episode took place in 278 BCE, when the Romans could no longer be sure who the god Summanus really was. All they knew was that the bolt of lightning that hit the statue was hurled by Jupiter himself[7]: the lunar cult was being replaced by a solar cult…

Notwithstanding this, the more or less magnificent spectacle of the Moon falling from on high continues to play itself out in the sky alternately every 3,396 days.

To conclude, one final observation: it goes without saying that it is impossible for a node to pass through the First Point of Aries every 20 June in every year of the Numan cycle; if this transit occurs on 20 June in year ten, then that is the only true feast to *Summanus*, and all other feasts are merely anniversaries of the occasion.

[6] Cicero, *de divinatione*, 1.16: *[…] cum Summanus in fastigio Iovis optumi maximi, qui tum erat fictilis, e caelo ictus esset nec usquam eius simulacri caput inveniretur, haruspices in Tiberim id depulsum esse dixerunt, idque inventum est eo loco, qui est ab haruspicibus demonstratus.*

[7] *Epitome*, Livy, 14.

Chapter 19
The Revolution of the Line of Apsides, and the Feast of *Compitalia*

The Moon's orbit around the Earth is elliptical. The Moon's distance from the Earth fluctuates roughly from a minimum of 356,500 km to a maximum of 406,500 km, and works out at an average of 381,500 km. The difference between minimum and maximum is around 6.5 % more or less than the average value. This difference is unaltered by the Moon's angular diameter, which varies from 29°30′ when the Moon is at its greatest distance from Earth, to 33°50′ when the Moon is at its closest.

Yet, at its closest, the Moon's surface appears to be 30 %[1] larger than its surface when it is at its most distant, as anybody may see with the naked eye from Earth (see Fig. 19.1).

The closest and furthest points on the Moon's orbit around the Earth are known as the lunar *perigee* and *apogee*.[2] The line that runs through these two points—the *line of apsides*—revolves in a counter-clockwise, or direct motion, direction (see Fig. 19.2) and completes a revolution with respect to the fixed stars in around 3,233 days, at a rate of +0.1114[3] degrees per day.

As the Moon also follows the same direct motion, it takes longer to return through the perigee—an *anomalistic month*—than it does to return to the same position with respect to the fixed stars—a *sidereal month*—and return through the same node—a *draconitic month*. An anomalistic month lasts 27.5546 days; a sidereal month lasts 27.3216 days; and a draconitic month lasts 27.2122 days (see Chap. 13 and Table 13.1).

---

[1] The square of 1.065 divided by the square of 0.935 is around 1.30: $1.065^2 \div 0.935^2 = 1.297$.

[2] *Perigee* and *apogee* are constantly varying—either increasing or decreasing—between the values of minimum perigee and maximum apogee. Yet "the extreme perigees and apogees take place only during the winter in the northern hemisphere, the period of the year when the Earth is closest to the Sun. For instance, all 14 closest perigees during a period from 1500 to 2500 CE fall between 6 December and 9 February", notes Meeus 1997, p. 15. He concludes: "It is evident that the Earth's variable distance from the Sun somewhat affects the Earth-Moon distance."

[3] 0.1114 = 360 ÷ 3,233.

L. Magini, *Stars, Myths and Rituals in Etruscan Rome*, Space and Society,
DOI 10.1007/978-3-319-07266-1_19

Fig. 19.1 The Moonat perigeeand apogee. At perigee, that is to say its minimum distance from the Earth, the Moon looks markedly larger and brighter than the Moon at apogee, when it is at its greatest distance. (from http://www.fourmilab.ch/earthview/moon_ap_per)

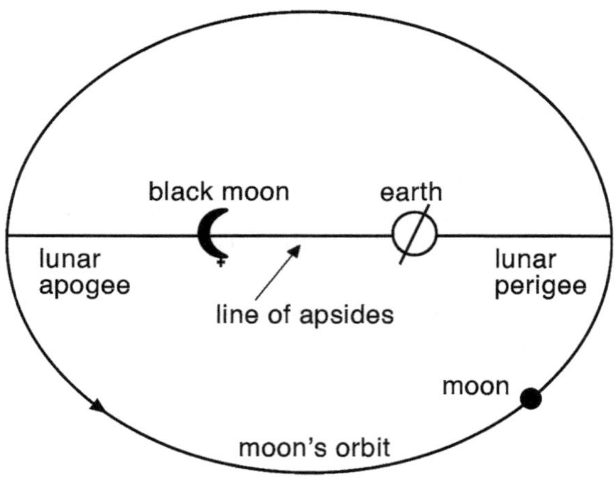

Fig. 19.2 The Moon's orbit around the Earth, the line of apsides and the Black Moon. The Moon's orbit is elliptical: the point closest to the Earth is the lunar perigee, and its farthest point its apogee. The line that connects these two points is known as the *line of apsides*, and it moves in an anticlockwise direction. The Earth occupies one of the two focuses on the ellipsis; the *Black Moon*—a fictitious heavenly body in astrology—occupies the other. "Viewed" from the Earth, the Black Moon is always near the lunar apogee, and moves in an anti-clockwise direction

The Moon's motion along its own orbit is not related to its position with respect to the Sun. In consequence, at apogee or perigee, the Moon might be in any phase: new, waxing, first quarter, full, last quarter, or waning. In its motion, however:

> The Moon appears to travel along the circle of the Zodiac in an irregular manner: if, to begin with, the Moon describes a certain arc, the following day it travels a wider arc, and in the days after that, it journeys along increasingly wider arcs, until it reaches its greatest arc; then the Moon describes increasingly smaller arcs, until it returns to the smallest arc from which it started. The interval of time between the minimum movement and the next minimum is known as an anomalistic revolution.

Geminus, a Greek astronomer from the first century BCE, thus describes a specific characteristic of the Moon's extremely complex motion. He explains:

> Astronomical observation shows that the *Exeligmos*[4] comprises 669 full months and 19,756 days. In this span of time, the Moon completes 717 anomalistic revolutions in longitude and, at the same time, completes 723 circles of the Zodiac plus 32°.[5]

One problem arising from these observations and data—all of which are of Mesopotamian origin, as the author himself notes—unnecessarily taxed the minds of Hellenistic Greece's greatest (and other) astronomers:

> What uniform and ordered movements may explain the apparent motion of the planets?[6]

Here we have an emblematic example of how prejudice trumps reason: who ever said that real movements should be "uniform and ordered", when the apparent motion—the "phenomena"—are variable and disordered?

Long before Renaissance astronomers in Europe—Kepler, to be precise—resolved the problem by explaining that the Moon's motion is irregular because it follows an elliptical rather than a circular orbit, Geminus tackled and resolved a smaller issue regarding the Moon's anomalous motion:

> With the assistance of these ancient observations on apparent motion, it was necessary to establish the Moon's daily longitudinal anomaly by calculating its minimum, maximum and average movements, along with its daily increase or decrease [...] The result is that the Moon's minimum movement amounts to 11°6′35″, its average to 13°10′35″, and its maximum to 15°14′35″; and that the daily increase is equivalent to 18′.[7]

The most interesting thing about Geminus's reasoning is that the Moon's "irregular" movement—this is what the Greek term "anomalous" refers to—has three possible progressions: minimum, average and maximum; a corollary in myth is not hard to find. Indeed, we must return to the 3,233 days of the revolution of the line of apsides. The 3,233rd day of the Numan cycle corresponds to 5 January[8] in year nine (see Tables 4.1 and S. C.). On this day, the line of apsides, together with

---

4 See Chap. 13.

5 Geminus, *Introduction to the Phenomena*, 18.2–18.3.

6 Simplicius, *Commentary on De Coelo*, 2.12.

7 Geminus, *Introduction to the Phenomena*, 18.4–18.5 and 19.

8 Please see again note 2.

the lunar perigee and apogee, returns to the same point with respect to the fixed stars as on the first day of the cycle.

That same day, 5 January, was the feast of *Compitalia* in Rome. "The crossroads", as it was known, was a *conceptiva* or moveable[9] feast, which celebrated a particular aspect of the Moon, represented in the heavens by *Hecate Trivia*, patron goddess of the crossroads.

It is beyond the scope of this book to investigate the enormous amount of material available on Hecate. For our purposes, it is sufficient to note that:

- The most ancient Western chronicler, Hesiod, defines her as the goddess

> that Zeus Cronides, of all the gods, honoured: he gave her splendid gifts, that she may have power over the Earth and over the shining sea: in the starry sky she too is honoured, revered in the highest by the immortals [...] In effect, of all those born to Gaia and Uranus worthy of honour, she has her part in all of their privileges[10];

- on occasion, Hecate is united with/confused with Selene, the "(full) Moon", with Artemis, the "(final crescent of the waning) Moon", and with Persephone, the "(new) Moon";
- she is known as *Polymorphes*, "of the many (different) forms, multiform";
- she is invoked as

> trioditis [...] heavenly, earthly and of the sea [...] the lady to whom the entire cosmos is entrusted.[11]

Ovid also highlights her triple nature with Janus's words:

> Thou seest Hecate's faces turned in three directions
> that she may guard the crossroads where they branch three several ways[12];

while Plutarch explicitly identifies Hecate with the Moon, explaining:

> Yet it is not with a single motion that she moves; but she is, as somewhere she is in fact called, "the goddess of three ways" [i.e. Trioditis or Trivia], for she moves on the Zodiac

[9] It is, of course, true that the moveable feasts in our own calendar, such as Easter, are associated with the motion of the Moon, while our fixed holidays, such as Christmas, are associated with the Sun's motion.

[10] Hesiod, *Theogony*, 411–415 and 421–422: *Ἡ δ' ὑποκυσαμένη Ἑκάτην τέκε, τὴν περὶ πάντων / Ζεὺς Κρονίδης τίμησε· πόρεν δέ οἱ ἀγλαὰ δῶρα, / μοῖραν ἔχειν γαίης τε καὶ ἀτρυγέτοιο θαλάσσης. / Ἡ δὲ καὶ ἀστερόεντος ἀπ' οὐρανοῦ ἔμμορε τιμῆς / ἀθανάτοις τε θεοῖσι τετιμένη ἐστὶ μάλιστα [...] / Οσσοι γὰρ Γαίης τε καὶ Οὐρανοῦ ἐξεγένοντο / καὶ τιμὴν ἔλαχον, τούτων ἔχει αἶσαν ἁπάντων.*

[11] *Orphic Hymns*, 1.1–1.2 and 7: *[...] τριοδιτιν, / [...] ουρανιαν χθονιαν τε και ειναλιαν, / [...] παντος κοσμου κληδουχον ανασσαν.*

[12] Ovid, *Fasti*, 1.141–1.142: *Ora vides Hecates in tres vertentia partes, / servet ut in ternas compita secta vias.*

against the signs in longitude and latitude and in depth at the same time. Of these movements the mathematicians call the first 'revolution,' the second 'spiral,' and the third, I know not why, 'anomaly'.[13]

Here we have a description of the three motions of the Moon—notably of the third, "depth", by which Plutarch means the Moon's distance from the Earth, its variations in the apogee and perigee—that reflects Hecate's three-faced appearance.

Hecate is invoked in a wide variety of occasions: Medea invokes her at midnight on the night of the full Moon, and then again at dawn; also at dawn, she is invoked by Aeneas and Dido; Demeter comes across her holding a torch aloft, an attribute that turns Hecate into *Phosphoros*, the "Bringer of light", just like the waning Moon or Venus at dawn (see Chap. 11, particularly note 6). Medea "for an entire day" looks after the Goddess's temple, and "at the first light of dawn" has herself borne there; when, that morning, she meets Jason, she pleads with him to attend for the "hour that sunders the night in two", for him to honour the Goddess and seek her protection for the trial that awaits him—a trial that begins at dawn and concludes at sunset.[14]

There are countless other examples we could add to these in order to demonstrate that Hecate does not personify a specific position or phase of the Moon in relation to the fixed stars or Sun. A more detailed and analytic investigation[15] would likely show that, since time immemorial, Hecate has been the ideal personification of the *Black Moon*, the Moon at its apogee. It is self-evident why this would be: of the great many appearances the Moon may assume, by far the most mysterious and disquieting is when it is at its apogee, when the changeable nature of the Moon's motion is compounded by its distance, diminished surface area and luminosity.

---

[13] Plutarch, *de facie*, 937E: *καιτοι μιαν ου κινειται κινησιν αλλ', ως που και λεγεται, Τριοδιτις εστιν, αμα μηκος επι του ζωδιακου και πλατος αντιφερομενη και βαθος. ον την μεν περιδρομην την δ'ελικα την δ'ουκ οιδα πως ανωμαλιαν ονομαζουσιν οι μαθηματικοι.*

[14] See: Ovid, *Metamorphosis*, 7.179–7.194; Valerius Flaccus, *Argonautica*, 7.331 and 335; Virgil, *Aeneid*, 6.247 and 255; 4.584–4.585 and 609; *Homeric Hymn to Demeter*, 52; Apollonius Rhodius, *Argonautica,* 3.251, 828 and 842; 956; 1029; 1224; 1407.

[15] In reality, the Goddess's sphere of competence is worthy of greater study (Hesiod, *Theogony*, 411–452), as is her relationship with the Lares, especially the *Lar familiaris* who, in the name of and on behalf of Vulcan, at the *Matralia* on 11 June conceived Servius Tullius with Ocresia (Ovid, *F.* 6.627–6.636); Servius Tullius was later to institute the *Compitalia* (Pliny, *Naturalis Historia*, 36.204; Dionysius, *Roman Antiquities*, 4.14.3; Macrobius, *Saturnalia*, 1.7.34) and, perhaps, establish the *Volcanal.*
As for Vulcan, his Greek counterpart Hephaestus was born to Hera without Zeus's intervention in response to Zeus's giving birth to Athena without Hera's intervention. Similarly, Athena represents the Moon in eclipse, whereas Hephaestus represents the Sun in eclipse. The mythical love between Hephaestus and Aphrodite reflects the fact that during a total solar eclipse, in full daylight the planet Venus is observable, inevitably in proximity to the Sun.

Indeed, in an astrological tradition whose roots are lost in the mists of time, the Black Moon that represents Hecate in the heavens becomes a truly unreal celestial body, located at the second focus of the elliptical orbit travelled by the Moon around the Earth. Since the Earth occupies the first focus, observers on Earth always see the Black Moon in the direction of the lunar apogee (see again Fig. 19.2); they will always see it revolving anticlockwise with the apogee, and completing its revolution in 3,233 days.

On the *Compitalia* feast that falls on 5 January in year nine of the Numan cycle, the line of apsides—and in particular the apogee, including the Black Moon—passes through the same longitude as it did 3,233 days earlier, on 1 March in year one. The comment at the end of Chap. 18 once again applies: as the line of apsides does not transit through a given longitude every 5 January in every year of the Numan cycle, if it makes the transit on 5 January in year nine, the only true feast to *Hecate Trivia*, goddess of the *Compitalia*, is that particular year; all others are mere anniversaries.

Chapter 20
Composition of the Revolutions of the Lines of Nodes and of Apsides

A revolution of the line of nodes with respect to the fixed stars takes 6,793 days (see Chap. 17); a revolution of the line of apsides with respect to the fixed stars takes 3,233 days (see Chap. 19). The former is retrograde, the latter is direct: their opposite motions add up one to the other.

The former travels at a rate of –0.0530 degrees per day; the latter at a rate of +0.1114 degrees per day.[1] The sum total is +0.1644 (= 0.0530 + 0.1114) degrees per day, and the two lines coincide again roughly every 2,190[2] days.

This value approximates to one quarter of the Numan cycle: 2,190 × 4 equals 8,760, an absolute error of –6.00 days and a percentage error of 0.068 % with respect to the 8,766 day cycle.

Though this may be *a hypothesis*, it once again demonstrates the solidity and elegance of the 24-year intercalation cycle (see Chap. 4), as it means that if the two lines begin moving from the same degree of longitude on day one of a Numan cycle, they will coincide a total of four further times within that particular cycle. Table 20.1 shows when these hypothetical encounters would occur over the 8,766-day cycle (see Tables 4.1 and S. C.).

Table 20.1 Composition of the revolutions of the lines of nodes and of apsides supposing that the two lines start at the same longitude on day one of the Numan cycle

Days of the cycle in which the lines of nodes and apsides coincide (rounded figures)	Corresponding days, months and years of the Numan cycle
1	1 March 01
2,190	20 intercalary 06
4,380	13 intercalary 12
6,570	28 February 18
8,760	21 intercalary 24

[1] I have extracted these figures—slightly rounded up/down—from Meeus 1997, pp. 11–14.

[2] 2,190 = 360 ÷ 0.1644.

L. Magini, *Stars, Myths and Rituals in Etruscan Rome*, Space and Society,
DOI 10.1007/978-3-319-07266-1_20

It may be noted that over the course of 2,190 days the line of nodes retrocedes by around 116°, while the line of apsides advances by some 244°[3]; it follows that each new coincidence takes place 244° further along. If we round this number down to 240°, after three encounters, i.e. after 720° (= 3 × 240°), the two lines will each have returned to their starting points. In other words, on 28 February of year eighteen (see Table 20.1) in the Numan cycle, they return approximately to the position where they began the cycle.

[3] –116.07° = 2,190 × –0.0530; + 243.97° = 2,190 × + 0.1114; obviously, 116° + 244° = 360°.

Chapter 21
The Numan Cycle as a Reference Framework for Celestial Motion

A comparison of celestial motion data with data on the Numan calendar and cycle proves that there is a concordance between the cadence of astral movements and the cadence of calendar feast days.

As we have seen, there is in particular a concordance with regard to four astronomical cycles that are crucial to describing the motion of the Earth, Moon and Sun. To recap, if these four cycles all commence simultaneously on 1 March in year one of the Numan cycle, at what can be defined as "the starting point of astronomical time" (see Chap. 1), we note that:

- the Saros cycle is completed 6,585 days later, on 15 March in year nineteen, on the feast of *Anna Perenna* (see Chaps. 13 and 14);
- the revolution of the line of nodes is completed 6,793 days later, on 14 October in year nineteen, the day before the feast of *October equus* (see Chap. 17);
- if one node—whether descending or ascending makes no difference—transits through the First Point of Aries on the first day of the Numan cycle, the other node—whether ascending or descending makes no difference—transits through this same point 3,396 days later, on 20 June in year ten, at the feast of *Summanus* (see Chap. 18);
- the revolution of the line of apsides is completed 3,233 days later, on 5 January of year nine, on the feast of *Compitalia* (see Chap. 19).

What we have here are four separate astronomical events that begin on the same day and end on four different days in four different years, all four of which next fall on feast days that recall myths and rites symbolizing those self-same astral events. One thing is certain: the myth of *Anna Perenna* describes an eclipse, while the *October equus* rite is based on a horse head and tail, and so on. However, the contrary does not apply: the *October equus* rite does not describe an eclipse, and the myth of *Anna Perenna* is not based on the *Head* and *Tail* of an animal that symbolizes the *Dragon*, and so on.

L. Magini, *Stars, Myths and Rituals in Etruscan Rome*, Space and Society,
DOI 10.1007/978-3-319-07266-1_21

What we must now do is ask ourselves what the probability is of these four concordances being a coincidence. Each one of them has a probability equal to $1 \div 355$ (one out of the 355 days in the Numan year) which works out at 2.8×10^{-3}, that is to say, a probability of 0.28 %. Given that the occurrence of each of these events is independent of the occurrence of the others, the probability that all four of these events take place corresponds to $(1 \div 355)^4$, or 6.3×10^{-11}, that is to say, a probability of 6 in 100 billion.

So, the probability of this being mere chance is vanishingly small. From here on in, we may assert that *the entire 8,766-day Numan cycle is a framework of reference* that not only covers the movement of the Earth, Moon and Sun; it covers the movement of the planets too.

We shall next be investigating how the Numan cycle serves the same function as what Mesopotamian astronomy referred to as the *Diaries* and *Goal Year Texts*.[1]

Available source material demonstrates that, from at least 747 BCE, the astronomical data contained in the *Diaries*

> were often estimated rather than measured. The date of solstices and equinoxes and of the heliacal rising of Sirius are recorded, but analysis has shown that at least during the Seleucid period (311 BCE onwards) these are calculated, not observed.

At the same time,

> the *Diaries* were used to compile a number of chronologically arranged tables of planetary and lunar phenomena. The tables of eclipse are of particular interest. The 18-year cycle of lunar eclipses and the succession of eclipse possibilities at 6 or occasionally 5-month intervals was probably already known to the Assyrians. Fragments of extensive tables of eclipses giving regnal year and month give an immediate overview of Babylonian chronology from 747 BCE onwards, and it may be this source material which enabled Ptolemy to compile a chronology of the Babylonian, Persian and Seleucid kings.

A feature of later *Diaries*, which date back to 236 BCE, are the *Goal Year Texts*:

> These exploit the known period relations between repeating phenomena of the Moon and the five planets: 71 years and 83 years for Jupiter, 8 years for Venus, 46 years for Mercury, 59 years for Saturn, 79 years and 47 years for Mars, and 18 years for the Moon [...]
>
> By looking at the *Diary* for the appropriate year in the past the scribe could compile a list of the phenomena to be expected in the year ahead (the *Goal Year*).

For each of the five planets, a whole series of relationships was known in Babylon, encompassing full multiples of synodic periods, full multiples of sidereal periods, and full multiples of solar years. Table 21.1 presents some of the multiples adopted in earlier and later *Goal Year Texts* (respectively GYT and GYT2).

For instance, Table 21.1 shows that over the course of 59 solar years, Saturn has 57 synodic phenomena of one form or another: heliacal risings, first stations,

[1] For more on Mesopotamian astronomy, see Britton-Walker 1996, from which I have extracted these quotes and the following data; see in particular pp. 50–5.

Table 21.1 Planetary period relations

Planet/Source	Synodic phenomena	Sidereal revolutions	Solar years
Saturn/GYT	57	2	59
Jupiter/GYT2	76	7	83
Mars/GYT2	37	5	79
Venus/GYT	5	3	8
Mercury/GYT	145	46	46

Table 21.2 Planetary period relations with the days, months and years of the Numan cycle in which each planet returns to its initial position

Planet	Solar years	Breakdown of solar years into Numan cycles and cycle years	Day, month and year in which the planet returns to its initial position with respect to the fixed stars and to the Earth
Saturn	59	=(24 × 2) + 11	28 February 11/III[a]
Jupiter	83	=(24 × 3) + 11	28 February 11/IV
Mars	79	=(24 × 3) + 7	4 March 08/IV
Venus	24 (=8[b] × 3)		28 February 24/I
Mercury	46	=24 + 22	1 March 23/II

[a] 24-year Numan cycles are indicated in Roman numerals: 11/III means the eleventh year of third Numan cycle

[b] Given that this period is shorter than the Numan cycle, for comparative purposes we have multiplied by three. As we saw in Chap. 9, no. 4, over the 24-year cycle, Venus has completed a total of 39 (=13 × 3) revolutions with respect to the fixed stars and of 15 (=5 × 3) revolutions with respect to the Earth

oppositions to the Sun, second stations or vespertine settings. More generally, in 59 solar years Saturn completes 57 synodic revolutions, that is to say, it returns 57 times to the same position with respect to the Earth; over the same period, it completes two sidereal revolutions, that is to say, it returns to the same position with respect to the fixed stars twice.

At the end of 59 solar years, Saturn has returned to its original position with respect to the Earth and to the fixed stars; in other words, it's as if it hasn't budged at all. If you are aware of Saturn's position with respect to the Earth and the stars over a 59-year period, you know where it has been and where it will be at all times. The same applies to Jupiter after 83 years, to Mars after 79 years (Fig. 21.1), and so on.

What we must do now is calculate what these multi-year periodical relationship cycles for the various planets correspond to in terms of Numan cycles (see Chap. 4 and Table 4.1). To recap, the Numan cycle lasts 24 years, is asymmetrical, and the first 11 years number 4,017 days, which is close to 11 solar years. We can now understand why that is: the multi-year cycles of two of the superior planets—Saturn and Jupiter (see Table 21.1)—are found by combining multiples of 24 with 11: respectively 59 = (24 × 2) + 11 and 83 = (24 × 3) + 11.

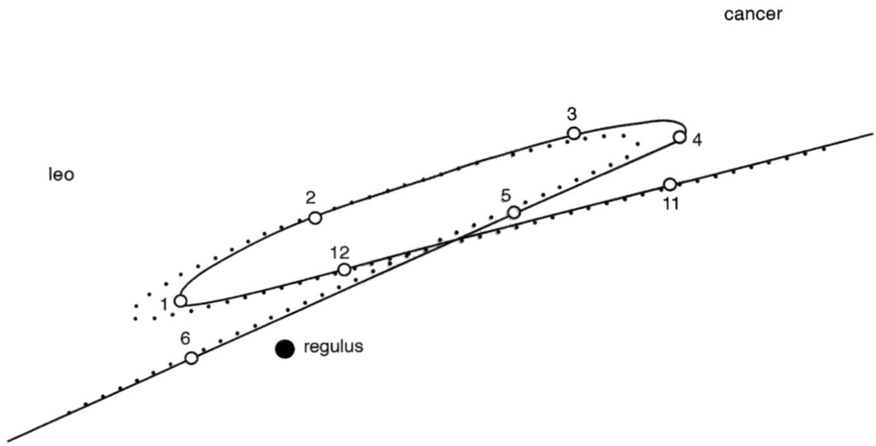

Fig. 21.1 Example of the periodical return of a superior planet to the same position with respect to the Sun and the fixed stars. Mars—here in retrograde motion near *Regulus*, the brightest star in Leo—during two passages 79 years apart: from November 1994 to June 1995, and from November 2073 to June 2074. (from Meeus 1997)

What we find is that Saturn returns to its initial position with respect to the Earth and the fixed stars at the end of two Numan cycles plus the first 11 years of the third, that is to say, on 1 March in year twelve of the third Numan cycle, or, if you prefer, day one of the 60th year. The same applies to Jupiter, at the end of three Numan cycles plus the first 11 years of the fourth, that is to say, on 1 March in year twelve of the fourth Numan cycle, or, if you prefer, day one of the 84th year.[2]

Two more things should be borne in mind with regard to Saturn and Jupiter. Table 21.1 shows that, for Saturn, 59 solar years correspond to 57 synodic periods for the planet: that's the same as saying that Saturn's synodic period lasts around 378 days,[3] in line with modern calculations; 378 days happens to be the length of the intercalated Numan year. Table 21.1 also shows that for Jupiter, 83 solar years correspond to 76 synodic periods for the planet: or, in other words, Jupiter's synodic period lasts around 399 days,[4] once again in line with modern calculations; given that 399 $\times$ 22 equals 8,778, it becomes clear that the Numan cycle is roughly equivalent to 22 of Jupiter's synodic periods.

Now, Mars returns to the position it was originally in on 1 March in year one on 4 March of year eight in the fourth cycle, that is to say, in the 80th year from the start.

---

[2] If the Numan cycle was divided into three equal parts lasting 2,922 days each, with an alternate intercalation of 20 and 21 days in even-numbered years (see Chap. 4, no. 8), Saturn and Jupiter would complete their multi-year cycles not on 1 March but on 10 March in year 12 of the third and fourth Numan cycles respectively.

[3] $59 \times 365.2422 = 21{,}549.289$; $21{,}549.289 \div 57 = 378.058$.

[4] $83 \times 365.2422 = 30{,}315.102$; $30{,}315.102 \div 76 = 398.88$.

Once again, the intercalation adopted fits well, considering that Mars is the closest planet to Earth, and the one with the most irregular motion.

For the planet Venus, the multi-year period is just eight years. However, if we multiply this by three, we find that the third time that the planet returns to its initial position with respect to the fixed stars and to the Earth coincides with the end of the 8,766-day Numan cycle, on 28 February in year 24.

Lastly, Mercury returns to its initial position after 46 solar years, or 16,801 days. Adding in the intercalation adopted (see Table 4.1), this brings us to 1 March in year 23 of the second Numan cycle.

Table 21.2 shows the relationship between planetary periods and Numan cycles for each planet. At the end of their respective time periods, all five planets have returned to their initial position with respect to the Earth and to the fixed stars—once again, it is as if they hadn't moved an inch. The one remaining problem is, with respect to the Earth and the fixed stars, how to locate the position of the planets within each individual interval; once we find this out, we know it for all time, in the past and the future. The place to look for the answer—if indeed it exists—to this is, not surprisingly, among the festivities of the Numan calendar.

Chapter 22
The Movements of the Superior Planets and the *Ancilia*, the "Sacred Shields"

The first feast day we shall consider is 1 March, day one, year one in the Numan cycle. At this starting point in astronomical time, Saturn, Jupiter, Mars, Venus and Mercury are all at a given position with respect to the Earth and the fixed stars—a yet-to-be-determined position; they are set to return to this position respectively 59, 83, 79, 24 and 46 years later.

In actual fact, 1 March is a feast day for two separate celebrations. The first regards Mars's descent to the Earth prior to the conception of Romulus, the act that marks the start date for calculations of astronomical time (see Chap. 1). The second regards another no-less-miraculous descent from the heavens.

Once again, Ovid provides arguably the best account of the story:

> Soft was the Earth with hoar frost spread like dew at morn,
> when the people gathered at the threshold of their king [Numa].
> Forth he came and sat him down in their midst upon a throne of maple wood;
> unnumbered men stood round him silent.
> Scarcely had Phoebus shown a rim above the horizon:
> their anxious minds with hope and fear did quake.
> The king took his stand, and, his head veiled in a snow-white hood,
> lifted up his hands, hands which the gods already knew so well.
> And thus he spoke: 'The time has come to receive
> the promised boon; fulfil thy promise, Jupiter.'
> Even while he spoke, the Sun had already lifted his full orb above the horizon,
> and a loud crash rang out from heaven's vault.
> Thrice did the god thunder from a cloudless sky, thrice did he hurl his bolts.
> Take my word for it: what I say is wonderful but true.
> At the zenith the sky began to yawn;
> the multitude and their leader lifted up their eyes.
> Lo, swaying gently in the light breeze, a shield
> fell down. The people sent up a shout that reached the stars.
> The king lifted from the ground the gift, but not till he had sacrificed a heifer,
> which had never submitted her neck to the burden of the yoke,

L. Magini, *Stars, Myths and Rituals in Etruscan Rome*, Space and Society,
DOI 10.1007/978-3-319-07266-1_22

and he called the shield *ancile*, because it was cut away (*re-cisum*) on all sides,
and there was no angle that you could mark.[1]

The astral backdrop to the scene is easy to surmise: the *ancile* falls in the short space of time (less than eight minutes) that it takes for the whole of the Sun's disc to cross the threshold of the horizon: "Scarcely had Phoebus (i.e. the Sun) shown a rim above the horizon [...] Even while he spoke, the Sun had already lifted his full orb above the horizon [...]" The signal is given by the loud crash ranging out, literally, "from heaven's vault"; the sky opens "at the zenith", traced by the augur's lithuus.

The framework of reference—or the coordinates, one might venture to say, anachronistically—of how the poet portrays the event regards the points known in astrology as the *Ascendant*, or Horoscope, and *Midheaven*. The *Ascendant* is the point where the Zodiac intersects with the Earth's horizon to the east; *Midheaven* is the meridian that unites north and south, passing through the zenith as seen from the point of observation. We also have the two diametrically opposite points, *Descendant* to the west and *Imum Coeli* at the nadir. These are the four "cardinal points"—as Manilius[2] calls them in his particularly detailed description—used in astrology for public and private horoscopes. If we wanted, we could provide a horoscope for the *ancile* that Jupiter gave the Romans.

Ovid has more to say on the matter:

Then, remembering that the fate of empire was bound up with it,
he [Numa] formed a very shrewd design.
He ordered that many shields should be made, wrought after the same pattern,
in order to deceive a traitor's eyes.
The work was finished by Mamurius; whether he was more perfect in character
or in smithcraft, it would hard for any man to say.
Bountiful Numa said to him, 'Ask a reward for your service.
If I have a reputation for honesty, you shall not ask in vain.'
He had already named the *Salii* from their dancing (*saltus*),
and had given them arms and a song to be sung to a certain tune.
Then Mamurius made answer thus: 'Give me glory for my reward,
and let my name be chanted at the end of the song.'

---

[1] Ovid, *Fasti*, 3.357-78: *Mollis erat tellus rorata mane pruina: / ante sui populus limina regis adest. / Prodit et in solio medius consedit acerno; / innumeri circa stantque silentque viri. / Ortus erat summo tantummodo margine Phoebus: / sollicitae mentesque speque metuque pavent. / Constitit atque caput niveo velatus amictu / iam bene dis notas sustulit ille manus / atque ita: 'Tempus adest promissi muneris', inquit; / 'pollicitam dictis, Iuppiter, adde fidem.' / Dum loquitur, totum iam sol emoverat orbem / et gravis aetherio venit ab axe fragor. / Ter tonuit sine nube deus, tria fulmina misit. / Credite dicenti: mira sed acta loquor. / A media caelum regione dehiscere coepit. / Summisere oculos cum duce turba suo. / Ecce levi scutum versatum leniter aura / decidit: a populo clamor ad astra venit. / Tollit humo munus caesa prius ille iuvenca, / quae dederat nulli colla premenda iugo, / idque ancile vocat, quod ab omni parte recisum est / quaque notes oculis, angulus omnis abest.*

[2] Manilius, *Astronomica*, 2.789: *cardines.* On *cardines*, see Manilius, *Astronomica*, 2.788-855.

> Hence the priests pay the reward that was promised
> or the work of old, and they invoke Mamurius.[3]

Punctilious as Ovid may have been, he fails to provide the two most telling details: the number of copies Mamurius made of the original ancile, and the number of the *Salii* that carried them in procession, singing, dancing and praising the ironsmith's craft. Fortunately, Plutarch steps into give us the number of copies made of the shield, that is, eleven, while other authors inform us that the priests who carried them numbered twentyfour.[4] These are the very same numbers—in a different arrangement—as the number of years that Saturn and Jupiter take to return to their initial position with respect to the Earth and the fixed stars (see Chap. 21 and Table 21.2).

There are further interesting particulars we learn about the *Salii*:

- the priests carry "a light Thracian shield on their left arm, an oblong circular shape with hollowed sides";
- their dances are led by a *praesul*, "who leaps around at the front", beating out the rhythm and showing which steps to perform, either *amptruare*, "moving around forward", or *redamptruare*, "moving around backward";
- their ceremonies, lasting 24 days from 1 to 24 March, include *mansiones*, "stopovers", where the priests consume sumptuous dinners and then spend the night;
- on 1, 9 and 23 March, the priests "move the *ancilia*"—they carry them to the sacrarium which only they may enter (perhaps the *Sacrarium Martis* in the *Regia*);
- the *Salii*'s songs are divided into *axamenta*, name lists of Gods and divine personages, and *versus*, songs of praise to the various divinities from whom they took their name (*Ianuli, Iunonii, Minervii*);
- all *Salii* are "protected by Jupiter, Mars, and Quirinus".

---

[3] Ovid, *Fasti*, 3.379-92: *Tum, memor imperii sortem consistere in illo, / consilium multae calliditatis init: / plura iubet fieri simili caelata figura, / error ut ante oculos insidiantis eat. / Mamurius (morum fabraene exactior artis, / difficile est ulli dicere) clausit opus. / Cui Numa munificus: 'Facti pete praemia', dixit; / 'si mea nota fides, inrita nulla petes.' / Iam dederat Saliis a saltu nomina ducta / armaque et ad certos verba canenda modos; / tum sic Mamurius: 'Merces mihi gloria detur / nominaque extremo carmine nostra sonent.' / Inde sacerdotes operi promissa vetusto / praemia persolvunt Mamuriumque vocant.*

[4] Plutarch, *Life of Numa*, 13; Dionysius, *Roman Antiquities*, 2.71; Livy, *ab Urbe condita*, 1.20, 1.27 e 5.52.

Fig. 22.1 The *ancilia* that the Salii carried in their processions. As Ovid and Plutarch explain, *ancilia* are neither round nor oval; they are "rounded off in all their parts, bereft of any angle". Drawing from sardonyx jewellery, provenance unknown, Republican age - Museo Archeologico Nazionale, Florence, Italy. (from A. Rich, *Dictionnaire des Antiquités romaines et grecques*, 1883). PD-OUD—Wikimedia Commons

Lastly, all of the songs are sung "in harmony with the number of months in the Italic year", and the *ancilia* are defined "shields for a single year."[5]

We may now understand why the shape of the shield that descended from the skies is so specific, and a particular on which all ancient chroniclers, from Ovid to Plutarch, insist. Indeed, in the passage we have just cited, Plutarch gives the following explanation of their name:

> The shields themselves are called *ancilia* from their shape; for this is not round, nor yet completely oval, like that of a standard shield, but has a curving indentation, the arms of which are bent back and united with each other at the top and bottom to make a 'curved', i.e. *ankylos*, shape.[6]

---

[5] On the shape of the buckler shield, Dionysius, *Roman Antiquities*, 2.71 and Varro, *de lingua latina*, 7.43; on the songs and dances, Festus, 334L; Paulus Festus, 7L. and 3L; on protection, Servius, *ad Aen.*, 8.663: *in tutela Iovis Martis Quirini*; on the relationship to months and years, Joannes Lydus, *de mensibus*, 4.2: [...] κατα τον των Ιταλικων μηνων αριθμον; *Liber Glossarum*, Cod. Vat. Palat. 1773 f. 40 v.: *scuta unius anni.*

As pointed out earlier, 9 March is the day on which the solar year that concludes the 23rd intercalary day begins (see Chap. 7).

[6] Plutarch, *Life of Numa*, 13: *Αὐτὰς δὲ τὰς πέλτας ἀγκύλια καλοῦσι διὰ τὸ σχῆμαῆ: κύκλος γὰρ οὐκ ἔστιν οὐδὲ ἀποδίδωσιν, ὡς πέλτη, τὴν περιφέρειαν, ἀλλ' ἐκτομὴν ἔχει γραμμῆς ἑλικοειδοῦς, ἧς αἱ κεραῖαι καμπὰς ἔχουσαι καὶ συνεπιστρέφουσαι τῇ πυκνότητι πρὸς ἀλλήλας ἀγκύλον τὸ σχῆμα ποιοῦσιν.*

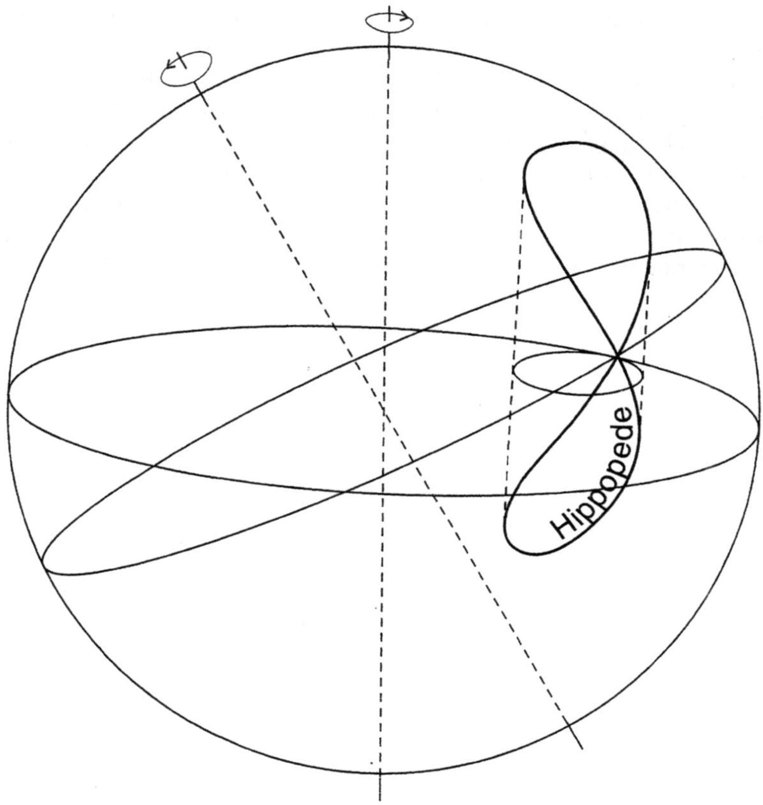

Fig. 22.2 The hippopede. The 8-shaped drawing represents a line traced on a spherical surface from a point on the equator of a sphere that spins at a constant speed; the sphere, in turn, revolves at the same speed around a different axis and in the opposite direction. (from Walker 1999)

The shape of the *ankylos* (Fig. 22.1) closely resembles that of the *hippopede*, a shape invented by Eudoxus of Cnidus in the fourth century BCE to explain the apparently irregular motion of the planets through a composition of uniform circular motions. The hippopede of each planet has a typical 8-shaped drawing, created by the composition of spherical planes tracing its synodic and sidereal periods (Fig. 22.2). However, the "8-shaped drawing" also resembles the *ancilia*.

In actual fact, the Greek term *ankylos* calls to mind the other Greek term *hippopede*: the former refers to a "form of attachment" or, more specifically, "a dog's leash" or "laces for shoes"; the latter originally referred to "shackles" or "a foot-tie for horses at pasture", and only took on its later astronomical reference figuratively.

Jupiter's gift to Numa, the *ancile* that fell from the sky on 1 March, resembles a prototype as-yet primitive tool for measuring multiples of the superior planets' synodic and sidereal periods. Measurement of these multiples commenced with the start of astronomical time, on 1 March in year one of the Numan cycle (see Chap. 1), and—as far as Saturn and Jupiter are concerned—concluded on 1 March of the 60 and 84th years respectively.

This confirms the entire cycle's status as a framework-providing cycle (see Chap. 21).

Chapter 23
The Heavens of the Fixed Stars and Their Custodian, *Ianus*

Mars, Jupiter and Saturn are the three superior planets visible to the naked eye, and man has known of them since the most ancient of times. Their sidereal periods lengthen the further they lie from Earth, from a little less than 2 years for Mars to almost 12 years for Jupiter and a touch under 30 years for Saturn. To the naked eye, the only thing beyond these planets is the sphere of fixed stars.

Mars, the closest of all, also has the most irregular motion: it is either impulsive and impetuous or slow, stationary or even retrograde. Little wonder that Mars was chosen as the heavenly representative of the god of war at least as far back as Babylon, when it symbolized the God Nergal. Jupiter is by far the most constant and balanced of the three planets, making it an ideal symbol for the most excellent and highest god whose calm wisdom guides and protects kings and populations alike. As for Saturn, with its exasperatingly slow pace and 30-year revolution, not far off the human lifespan in ancient times—or if you prefer, a "generation" in the ancient system for measuring long periods of time—it is better suited to the role of a being whose long scythe harvests the fruits of a lifetime's of work.

As is generally known, the farthest-flung gods—gods grafted onto earlier incarnations from the East, represented in the heavens by these relaxed and solemn planets—are united and split by family relations and conflictual relationships, including over Rome. First, Saturn fled to Rome to escape Jupiter; there he was greeted by Janus, who gave him the Capitol and then withdrew to the Janiculum; subsequently, Jupiter arrived, claimed the Capitol and confined Saturn to the foot of the hill (see again Fig. 16.1).

It is highly likely, if not certain, that this myth echoes not just earlier Mesopotamian notions of the heavens, but also traces of a preliminary sketch of another motif: that of the "great conjunction". The conjunction of the two largest planets may occur every 20 years, but it takes 24 centuries for the grand conjunction to travel the entire way around the Zodiac and repeat in the same sign, in the same

L. Magini, *Stars, Myths and Rituals in Etruscan Rome*, Space and Society,
DOI 10.1007/978-3-319-07266-1_23

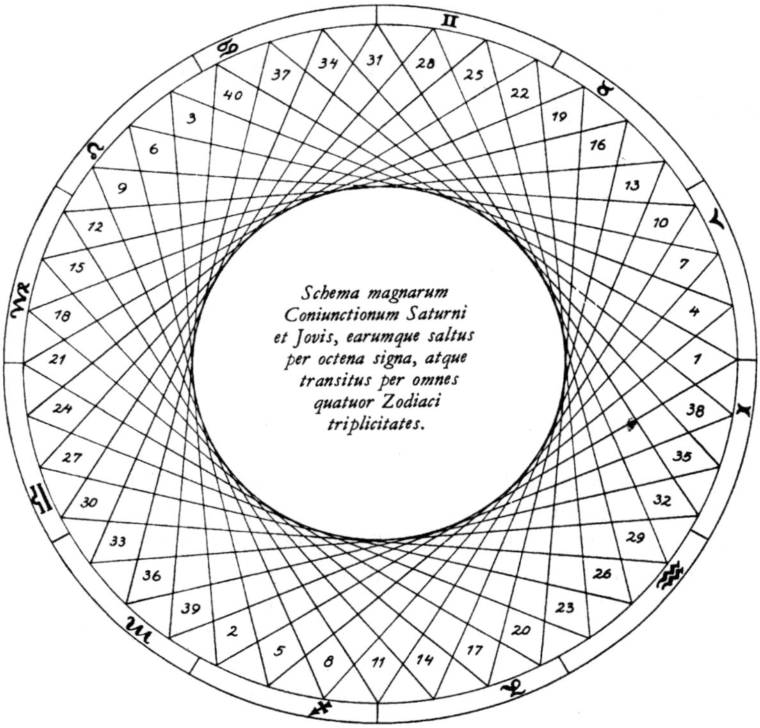

Fig. 23.1 The "great conjunction" of Saturn and Jupiter—from Johannes Kepler, *De Stella Nova in Pede Serpentarii*, 1604. This diagram shows the trigon formed from the "major conjunctions" of Jupiter and Saturn, which are repeated every 20 years; one angle of the trigon takes around 2,400 years to complete its journey around the Zodiac

position, as it occurred all that time ago (see Fig. 23.1). If there is one astral configuration that connotes the passage of time, and indeed the passage of entire epochs, this is a prime candidate. Clearly, the age of Saturn that preceded the age of Jupiter has always been regarded as the one true golden age.

The bond between Saturn and Janus in Rome takes on a specific value when read against this hastily-sketched framework. Ovid tells us that a god like the two-faced deity was Roman alone, "for Greece hath no divinity like thee".[1] Joannes Lydus grasps the meaning of this, when on Varro, he writes:

> In the XIV book of the *De Rerum Divinarum*, Varro [writes] that *Janus* for the Etruscans was known as *Uranus*, who looked over everything they did [...] In his book *De Signis*,

[1] Ovid, *Fasti*, 1.90: *nam tibi par nullum Graecia numen habet.*

> Fronteius [writes] that Janus was believed to watch over all time, and that his temple contained twelve altars, one for every month.[2]

"Janus was known as *Uranus*", and Uranus was the divinity associated with the high heavens, the heavens of the fixed stars. The bond between Janus and Saturn—a bond that the earliest Roman hagiography situated in the era before the city was founded—reveals the tenor of the relationship that unites the heavens of the fixed stars and the planet whose orbit is closest to them.

Ovid's description of Janus is more of an illustration, as may be inferred from his writings on the God:

> Two-headed Janus, opener of the softly gliding year,
> thou who alone of the celestials dost behold thy back.
> [...] alone of all the heavenly ones
> thou doest see both back and front.

The poet continues, leaving the god speak:

> 'The guardianship of this vast universe is in my hands alone,
> and none but me may rule the wheeling pole.
> [...] I sit at heaven's gate with the gentle Hours;
> my office regulates the goings and the comings of Jupiter himself.
> [...] I, the porter of the heavenly court, behold at once both East and West.
> [...] lest I should lose time by twisting my neck,
> I am free to look both ways without budging.'[3]

In this last quote, Ovid clearly contrasts the immobility of the skies where the stars reside—indeed, they are known as fixed stars—and the rest of the heavens, in which, as they gradually become lower and lower and closer and closer to Earth, the planets come and go, starting with Saturn. The planets make their appearance in the east, move in direct motion, appear to stand still, retrocede, go into opposition with the Sun, appear to stand still once more, resume direct motion, and then exit stage west.

All of these movements and phenomena are alien to Janus's nature—Janus, "thou who alone of the celestials", "alone of all the heavenly ones", "opener of the softly gliding year"—yet they are all movements and phenomena that the guardian of this vast world, "porter of the heavenly court", must keep under observation without turning his head and without moving his body.

---

[2] Joannes Lydus, *de mensibus*, 4.2: *ο δε Βαρρων εν τη τεσσαρεσκαιδεκατη των θειω νπραγματων φησιν αυτον παρα Θουσκοις ουρανον λεγεσθαι και εφορον πασης πραξεως [...] Φωντηιος δε εν τω περι αγαλματων εφοροναυτον οιεται του παντος χρονου τυγχανειν, και ταυτη δωδεκαβωμον ειναι τον αυτου ναον κατα τον των μηνων αριθμον.*

[3] Ovid, *Fasti*, 1.65-6, 91-2, 119-20, 125-6, 139-40, 143-4: *Iane biceps, anni tacite labentis origo, / solus de superis qui tua terga vides.*; *[...] de caelestibus unus / sitque quod a tergo, sitque quod ante vides.*; *'[...] me penes est unum vasti custodia mundi / et ius vertendi cardinis omne meum est.*; *[...] praesideo foribus caeli cum mitibus Horis: / it, redit officio Iuppiter ipse meo.*; *[...] ego perspicio caelestis ianitor aulae / Eoas partes Hesperiasque simul.*; *[...] et mihi, ne flexu cervicis tempora perdam, / cernere non moto corpore bina licet.'*

He must, at one and the same time, be *Patulcius* and *Clusius*, "He who opens" and "He who closes":

'[...] for on his sacrificial lips
I'm now *Patulcius* and now *Clusius* called.'[4]

This is, of course, a pre-scientific explanation of a pre-scientific Zero Degrees longitude. Whether it is the Greenwich Meridian (which measures longitude on the earthly sphere) or is the Zero Degrees of Aries (which measures longitude on the heavenly sphere), in all cases what we have is a point in space from which things begin and end, "open" and "close", and from where it is possible to measure the size of an angle.

Moreover, it is

> the statue of the god, more often than not depicted holding the number 300 in his right hand and the number 65 in his left, illustrating the length of the year,[5]

that gives the measurement in days of Janus's domains.

One key and final proof of the affinity between Janus and the vault of heaven is to be found in a comparison between an element of the Mesopotamian view of the cosmology and a myth dating back to the earliest times of Ancient Rome. The Near Eastern element is best summed up in the biblical cosmology (see Fig. 23.2), according to which,

> the earth rests on pillars (*Job* 9.6). Stretched above the earth is the sky, the 'heaven' or the 'firmament', a solid substance (*Genesis* 1.6-8) also resting on pillars (*Job* 26.11). The sun, moon and stars are positioned in, or just beneath, the firmament (*Genesis* 1.14-7) and they move across it (*Psalms* 19.1-7) [...]
> There are waters above the firmament (*Genesis* 1.6-7) as well as beneath it. Some of the waters beneath the firmament were gathered together at the beginning of creation to form the seas (*Genesis* 1.9-10), but, in addition, these waters flow beneath the earth (*Esodo* 20.4; *Deuteronomy* 4.18; *Psalms* 24.2) where they are connected to the waters of *Tehom*, the great deep (*Genesis* 1.2). The Deluge was caused by a tremendous outpouring of the fountains of the *Tehom* as well as by the opening of the windows of heaven (*Genesis* 7.11) [...]
> Rain is produced by the clouds (*Genesis* 9.11-7; *Job* 26.8; *Eccl.* 11.3). The water in the clouds comes from the waters above the firmament so that when the heaven is 'shut up' there is no rain (*Deuteronomy* 11.17), while when the 'good treasure' of heaven is open the rain falls in abundance (*Deuteronomy* 28.12).[6]

Macrobius also retells the Roman myth:

> During the war against the Sabines, provoked by the rape of the virgins, the Romans wanted to move quickly and close the gate at the foot of the Viminal Hill – thereafter

---

[4] Ovid, *Fasti*, 1.129-30: *[...] modo namque Patulcius idem / et modo sacrifico Clusius ore vocor.*

[5] Macrobius, *Saturnalia*, 1.9.10: *[...] simulacrum eius plerumque fingitur manu dextera trecentorum et sinistra sexaginta et quinque numerum tenens ad demonstrandam anni dimensionem.*

[6] See Louis Jacobs, 'Jewish Cosmology', in Carmen Blacker and Michael Lowe (eds.), *Ancient Cosmologies*, London 1975, pp. 66–86, from Internet. See also Jacobs 1978, pp. 55–57.

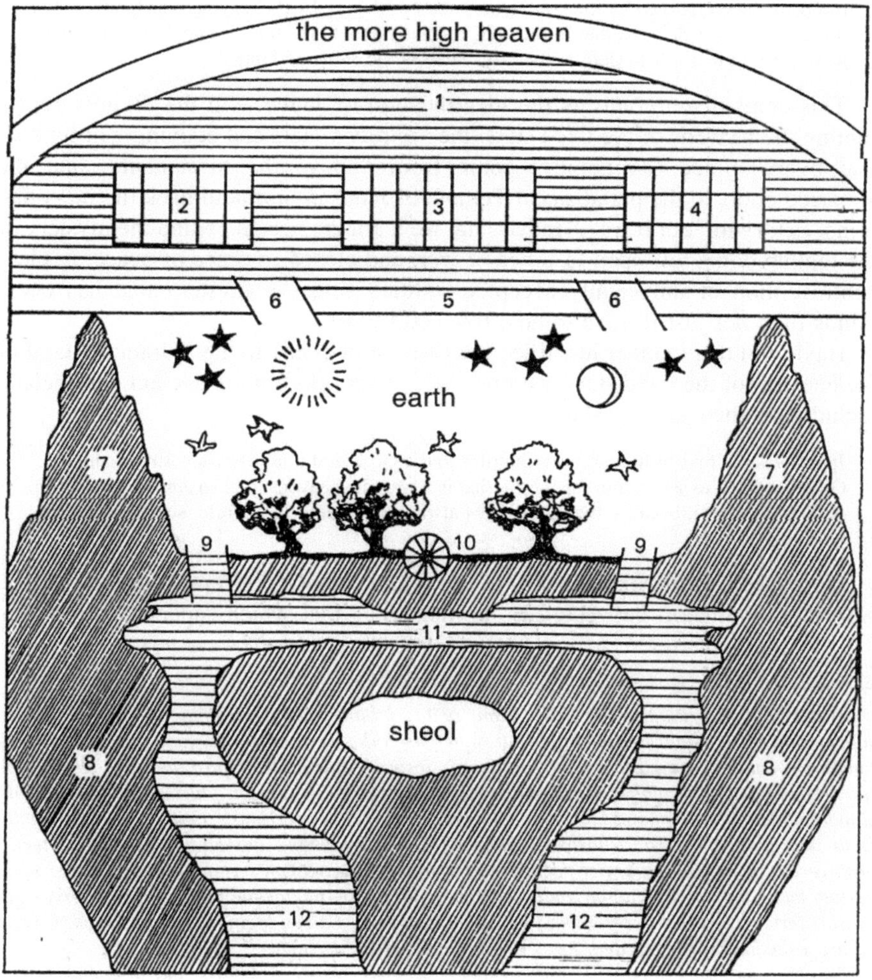

Fig. 23.2 The Mesopotamian cosmological vision in biblical interpretation: (*1*) water above the firmament; (*2*) deposits of snow; (*3*) deposits of hail; (*4*) halls of the winds; (*5*) firmament; (*6*) locks; (*7*) pillars of heaven; (*8*) pillars of the Earth; (*9*) fountain of the abysses; (*10*) centre of the world; (*11*) waters below the earth; (*12*) rivers of the underworld. (from Jacobs 1978)

known as the *Janual* – because their enemies had launched their assault there. As soon as they closed it, it opened again, all by itself; the same thing happened a second and a third time. As it was impossible to close the gate, a large number of armed guards were placed beyond the threshold. While fierce fighting was going on elsewhere, a rumour spread that our soldiers had been routed by Tatius. On hearing this news, the Romans who were defending the gateway fled in panic.

However, just as the Sabines were about to pour through the open gate, it is said that from Janus's Temple rushing and unruly waters flowed out through the gate, and many enemy troops perished, burned by the boiling surge or swallowed by its eddying whirlpools. After this event, it was decreed that in times of war the doors to the temple should remain open, as if the god had left to defend the city. Such is the story of Janus.[7]

This comparison confirms the identification of Janus with the divinity representing the heavens of the fixed stars: the "unruly waters and eddying whirlpools" that overwhelmed the Sabines come from "above the firmament", and the *un*barred doors to the paradisiacal Temple of Janus *do not* hold back the rains that bring victory to "our boys".[8] In passing, we learn the reason behind the mystery of the temple doors left open in times of war, and closed in times of peace, in clear contravention of universally accepted military rules—a mystery that has taxed minds (and not just Roman minds) for 3.000 years…

Having in this manner identified the essence of *Janus* in the "Uranian" God of the heavens of the fixed stars, we find more supporters among ancient chroniclers, including—once again—Macrobius:

Janus was the first in Italy to raise temples to the Gods and establish rites for worship […] Others wanted to see in him the world, that is to say the heavens: and so *Janus* comes from *ire* [i.e. "to go"], because the world goes always, travelling in a circle, starting from and returning to itself.[9]

If we take a step backwards and return to the Oriental conception of the cosmos, we see a distinction between the "firmament"—that which separates the waters

---

[7] Macrobius, *Saturnalia*, 1.9.17-8: *Cum bello sabino, quod virginum raptarum gratia commissum est, Romani portam quae sub radicibus collis Viminalis erat, quae postea ex eventu Ianualis vocata est, claudere festinarent, quia in ipsam hostes ruebant, post quam est causa, mox sponte patefacta est; cumque iterum ac tertio idem contigisset, armati plurimi pro limine, quia claudere nequibant, custodes steterunt, cumque ex alia parte acerrimo proelio certaretur, subito fama pertulit fusos a Tatio nostros. Quam ob causam Romani, qui aditum tuebantur, territi profugerunt. Cumque Sabini per portam patentem inrupturi essent, fertur ex aede Iani per hanc portam magnam vim torrentium undis scatentibus erupisse, multasque perduellium catervas aut exustas ferventi aut devoratas rapida voragine deperisse. Ea re placitum ut belli tempore, velut ad urbis auxilium profecto deo, fores reserarentur. Haec de Iano.*

The myth is also recounted, among others, by Ovid, *Fasti*, 1.255-82 and *Metamorphoses*, 14.772-804.

[8] Physically, the springs from whose waters the Sabines were routed are the *Lautulae*, so called—as Varro, *de lingua latina*, 5.156, explains—"from *lavare*, "to wash", because there [a little to the east of the *Comitium*; author's note] near the Double Janus there once were hot springs, *ab lavando, quod ibi ad Ianum Geminum aquae caldae fuerunt*." The temple of *Ianus Geminus* (see again Fig. 16.1) should be located midway between the *niger lapis* and the temple of *Cloacina*.

Etrusco-Roman tradition also mirrors Mesopotamian cosmology in other correspondences between the God and spring waters bubbling up from underground. One such link is the *lucus Furrinae*, the "sacred wood of Furrinae", the nymph of springs and divinity of the waters that emerge half way up the Janiculum, the hill named after the god; see Dumézil 1975, pp. 32–37.

[9] Macrobius, *Saturnalia*, 1.9.3 and 11: *[…] Ianum in Italia primum dis templa ferisse et ritus istituisse sacrorum […] Alii mundum, id est caelum, esse voluerunt, Ianumque ab eundo dictum, quod mundus sempre eat, dum in orbem volvitur et ex se initium facies in se refertur.* Arnobius (*adversus nationes*, 3.29.3), though disputing it, restates the opinion that Janus is the world, the year or the Sun.

below from the waters above, which God calls the heavens and created on the second day—and the "lights" in the firmament of the heavens (distinguishing day from night), which act as symbols for the seasons, the days and the years, as well as illuminating the Earth: i.e. the Sun, the Moon and the stars, created by God on the fourth day.

The term "firmament" references a notion of "firmness, closure"[10] and is used to define the "wrapper" around the various heavens, from the outermost heaven of the fixed stars to the innermost heaven of the Moon, at the centre of which sits the unmoving Earth. In turn, the concept of "firmness" hails back to the Sanskrit adjective *dhruvah*, meaning "firm, fixed, constant", which, by a fact of nature, gives its name to the Pole Star, the most fixed of all fixed stars—indeed, the one true fixed star which never moves, the Pivot Point of the Heavens.

It is from the Pole Star above the heads of Northern hemisphere dwellers that the Axis of the World turns, and it is around this axis that the entire universe revolves. Though this axis connects with the opposite pole, we cannot see it because it runs below our feet. It runs through the Earth, passing through an *Umbilicus* which, evidently, sits right before the observer, and moves when the viewer moves. This explains, more generally, the infinite number of *Umbilici* through which an infinite number of Axes pass in this world of infinite places sacred to human beings and, more specifically, the contiguity of the *Umbilicus Urbis* with the temple of *Janus Geminus* in Rome.[11]

So "universal" is this approach—no term could be more appropriate to define it—that we find it in all ages among all peoples, though of course with variations. Often, the Axis of the World is a tree whose roots tap into an *Umbilicus* of the World, serving as a pillar to hold up the celestial vault; in other cases, it is a mill that revolves around the axis[12]; in yet others—for example, in India—Heavens and Earth "are compared to two wheels placed at either end of an axle."[13] In all cases there is a piece of wood, sometimes even a little wooden rod—a cork on the world's belly button, or a wooden pintle fastening the wheel to the mill or the two wheels to the axis—which does not move, does not rotate, is fixed, and whose mere presence ensures eternal motion for the mill, wheels, heavenly bodies and universe.[14] The Sanskrit term *anih* would appear to resemble the Etruscan name for Janus, *Ani.*

In Classic Greece, under the Pythagorean tradition, whose roots are sunk into the Near East, this vision of the Cosmos takes on a different and highly particular

---

[10] Note the Latin expression *firma ianua*, "closed door", and the French verb *fermer*, "to close".

[11] See Coarelli 1983, p. 199 and ff., which suggests identifying the *Umbilicus* with the *Mundus*; see again Fig. 16.1.

[12] On the Axis of the World, see de Santillana-von Dechend 1983, *passim.*

[13] Gombrich 1978, p. 93.

[14] In Sanskrit, this diminutive but essential little wooden item has a name, *anih*, designating "the pintle inserted into an axle on the outside of a wheel to prevent it from slipping off". Hymn 1.35 in the *Rig Veda*, stanza 6, runs thus: "The immortal things rest upon this (*scil.*: the heavens of Yama, which is the "house of the heroes"), like a cart wheel on its pintle (*anih*)."

Fig. 23.3 The oldest known Roman coin, fourth century BCE. On the obverse, Janus is looking in two directions; on the reverse, a ship's bow. Bronze, ca. 225 BCE, 6 cm in diameter. *Left*: Cabinet des Médailles, Paris—*Right*: Pergamonmuseum, Berlin. Photo: Siren-Com and O. Mustafin—Wikimedia Commons

cant. The Pythagorean Philolaus of Croton, and later on Plato himself, mention a "hull", something that "holds together the entire revolving firmament, much like the bindings round a warship's keel."[15]

The two-way facing god's dual heads and the "hull", the "ship's keel", appear on the two sides of the oldest known Roman coin: the former, on the front, recalls that "lest he should lose time by twisting his neck, he is free to look both ways without budging" his head; the latter, on the reverse, symbolizes his function as the immobile "binding" of the universe (see Fig. 23.3).

[15] For Philolaus, who uses the terms *olkas*, "hull of a cargo ship", see *I Presocratici* (edited by Lami 1991), A15 (Aetius 2.6.5) pp. 476–477 and B12 (Theon), pp. 496–497; for Plato, *Republic*, 10.621d. See de Santillana-von Dechend 1983, p. 279 and ff.

Chapter 24
From the *Robigalia* on April 25 to the Entrance of the Sun in Aries for the Festivity of *Anna Perenna* on March 15

During the course of the so-called "Calendar of agricultural tasks", Pliny talks at length about the relationship between the days of the year, the rising or setting of the constellations, and the tasks to be carried out in the fields. One of these is particularly interesting, because it refers to the Numan calendar in force prior to Julius Caesar's reforms, and indicates the Sun's position in the Zodiac on a given day under that particular calendar:

> The *Robigalia* were instituted by Numa in the 11th year of his reign [705 BCE], and they are *now* celebrated on the 7th day before the Calends of May [25 April], as it is more or less in this period that blight afflicts the harvest. Varro set this date based on the position of the Sun in the 10th degree of Taurus, according to calculations used *in those times*.[1]

Under the Numan calendar, April lasts 29 rather than 30 days (as it does in the Julian calendar), and 25 April is the 6th rather than the 7th day before the Calends of May. So, there can be no doubt that the counter position of the *nunc* of the celebration and the *tunc* of the calculation period refers to the "now" of when Pliny was writing, subsequent to the Julian reform, and the "then", i.e. "in those times", when Varro made his calculations, prior to the reform.

What is more open to question is the Sun's position on 25 April in year eleven in the reign of Numa, which is traditionally held to be 705 BCE. Unfortunately, this identification is hypothetical, and there remains no way for us to associate the day, month and year of ancient Roman chronology with the day, month and year of our current chronology.

However, one thing we do know is that around 48 BCE—the year in which Varro wrote one of his many lost works, the *Antiquitates Rerum Divinarum*—on 25 April the Sun was at 10° in Taurus. With this information, it is not hard to

[1] Pliny, *Naturalis Historia*, 18.285: *Robigalia Numa constituit anno regni sui XI, quae nunc aguntur a. d. VII kal. Mai., quoniam tunc fere segetes robigo occupat. Hoc tempus Varro determinavit sole tauri partem X obtinente, sicut tunc ferebat ratio.*

L. Magini, *Stars, Myths and Rituals in Etruscan Rome*, Space and Society,
DOI 10.1007/978-3-319-07266-1_24

calculate the Sun's entry into the sign of Aries. We may choose between the following two methods:

- Attribute constant motion to the Sun over the entire year, equivalent to 0.9856° per day;[2]
- Attribute an accelerated motion to the Sun—equivalent to 1.0159° per day—from 13° in Virgo to 27° in Pisces, and a slower motion—equivalent to 0.9524° per day—from 27° in Pisces to 13° in Virgo, in accordance with the convention in use in Babylon.[3]

Over such a short arc of a circle—the 40° from Degree Zero of Aries to 10° in Taurus—the two methods yield just a single day's difference, between 41 days calculated using the first method[4] and 42 days using the second.[5] Far more striking is that the second method yields a round result to the third decimal place and, more important still, the 42nd day before 25 April is 15 March, the feast day of *Anna Perenna* (see Chap. 14). Ultimately Pliny tells us that on 25 April the Sun is at 10° in Taurus; a calculation using the Babylonian method indicates that on 15 March, the feast of *Anna Perenna*, the Sun is at 1° in Aries.

Among Ovid's many tales and writings on the Goddess, we find a match with Themis—"Others deem that she is Themis"[6]—wife to Zeus and mother of the Horae and Moirae. The Horae personified the seasons, while the Moirae mapped out the portion of the Zodiac that allocates to each man, according to the time and place of their conception and/or birth, the fate written for them in the stars and dictated by the divine will that Themis herself represents.

The pseudo-etymology of the Goddess's name advanced by Ovid takes on a new meaning:

> In a perennial river I hide, and Anna Perenna is my name.[7]

The eternal flow of the stars across the heavens recalls the continuous flow of the brook where one of the goddess's personifications found shelter; at the same time, the circle of the Zodiac continues to open and close at the very same point, Degree Zero in Aries.

---

[2] 0.9856° = 360° ÷ 365.25.

[3] The more accelerated motion, equal to 30° per lunation, yields 1.0159° (= 30° ÷ 29.5306) per day. The less accelerated motion, corresponding to 28.125° (= 15/16 × 30°) per lunation, yields 0.9524° (= 28.125° ÷ 29.5306) per day. See Pannekoek 1989, p. 73.

[4] This is the rounded result of 40.5844 = 40° ÷ 0.9856°.

[5] Over the arc of the Zodiac, considering that the Sun's is the slowest motion. The rounded result of 41.99916 = 40° ÷ 0.9524° is, precisely, 42.

[6] Ovid, *Fasti*, 3.658: *Pars Themim […] putat esse*. For Themis, Zeus, the Horae and the Moirae, see Hesiod, *Th.*, 901–906.

[7] Ovid, *Fasti*, 3.654: *Amne perenne latens Anna Perenna vocor.*

Modern scholars who consider the 15 March celebration "in all likelihood to be the start of an extremely ancient year",[8] would appear to be on the right track. Macrobius certainly agrees:

> That same month [i.e. March] people make a sacrifice in public and in private to Anna Perenna, so that the coming year and future years will flow pleasantly by.[9]

Only the truly ancient year that starts on 15 March, the feast of *Anna Perenna*, qualifies as a "Zodiacal New Year", with the Sun entering Aries.

[8] See Della Corte 1969, p. 112.

[9] Macrobius, *Saturnalia*, 1.12.6: *Eodem quoque mense et publice et privatim ad Annam Perennam sacrificatum itur, ut annare perennareque commode liceat.*

Chapter 25
From the Entrance of the Sun in Aries to Its Entrance in Pisces for the *Terminalia* on February 23, together with the Vespertine Rise of Arcturus

If we now repeat the twin process, starting from the Sun's entrance into Aries on 15 March, to calculate the day on which the Sun has travelled 330° and is entering into the sign of Pisces, we obtain two new results which are very close together. This is perhaps not surprising, as the distance between the start point and arrival is 330° in an anticlockwise direction; clockwise, the difference is just 30°. The first method gives us 335 days,[1] and the second 336.[2] Once again, the most exciting result is the second one: 336 days after the *Anna Perenna* celebration of 15 March comes *Terminalia* on 23 February.

The Sun's entrance into the sign of Pisces for the *Terminalia* on February 23 is significant for a number of reasons. Pliny, for one, writes that;

> Variable weather is expected with the appearance of swallows the 8th day before the Calends of March [22 February], and the day after at the vespertine rise of Arcturus.[3]

The arrival of the swallows is confirmed by Ovid, who on 24 February writes:

> Do I err? or has the swallow come, the harbinger of spring,
> and does she not fear lest winter should turn and come again?[4]

---

[1] Rounding 334.82 = 330° ÷ 0°,9856.

[2] Rounding 335.533 = 171.147 + 164.386; 171.147 = 163° ÷ 0.9524°; 164.386 = 167° ÷ 1.0159°.

[3] Pliny, *Natural History*, 18.237: *varie et VIII kal. Mart. hirundinis visu et posteriore die arcturi exortu vespertino.*
The evening rise—or vespertine rise—of a celestial body is the moment of the year when the star begins to be visible to the East soon after the Sun sets in the West. The morning rise is the moment of the year when the star begins to be visible to the East soon before the Sun rise in the East.
Hesiod too comments on the arrival of swallows at this time of year contemporaneous to the evening rise of Arcturus; see Chap. 26.

[4] Ovid, *Fasti*, 2.853–4: *Fallimur, an veris praenuntia venit hirundo / et metuit ne qua versa recurrat hiems?*

L. Magini, *Stars, Myths and Rituals in Etruscan Rome*, Space and Society,
DOI 10.1007/978-3-319-07266-1_25

This is the day when the God *Terminus* is celebrated, as Ovid explains:

What happened when the new Capitol was being built? Why, the whole company
of gods withdrew before Jupiter and made room for him;
but Terminus, as the ancients relate, remained where he was found
in the shrine, and shares the temple with great Jupiter.
Even to this day there is a small hole in the roof of the temple,
that he may see naught above him but the stars.[5]

So, there is somebody who observes the stars in Rome, and who is keen to continue observing them. This was prior to 583 BCE, the year when the first Etruscan king, Tarquinius Priscus (616–579 BCE), began construction of the new temple dedicated to the Capitoline triad Jupiter, Juno and Minerva.[6] The stars they wished to observe—perhaps the vespertine rising of Arcturus that Pliny mentioned?—and whether they wanted to observe them solely on 23 February is something we have yet to establish (see Chap. 27).

What we do know is that in Rome, between the eighth and sixth centuries BCE:

- Firstly, Arcturus (Fig. 25.1), the brightest star in the northern hemisphere, rises with the azimuth close to 45° (Fig. 25.2), in the North-East,[7] and sets with the azimuth close to 315°, in the North-West;
- and secondly, the temple of Jupiter Capitoline is oriented with the azimuth at 157° (Fig. 25.3), facing South-SouthWest.[8]

It follows that these three alignments—the rising and setting of the celestial body and the Capitoline Temple—respect the division of the circle of the horizon into sixteen parts under the so-called *Disciplina etrusca*:[9] the heavenly body rises in the direction that separates the second and third portion and sets in the direction that separates the 14 and 15th portion, while the temple is oriented in the direction separating the seventh and eighth portion (see again Fig. 25.2).

[5] Ovid, *Fasti*, 2.667–72: *Quid, nova cum fierent Capitolia? Nempe deorum / cuncta Iovi cessat turba locumque dedit; / Terminus, ut veteres memorant, inventus in aede / restitit et magno cum Iove templa tenet. / Nunc quoque, se supra ne quid nisi sidera cernat, / exiguum templi tecta foramen habent.*
Terminus's refusal to shift the altar dedicated to him to make space for the new temple is chronicled in Livy, *ab Urbe condita*, 1.55 and Dionysius, *Roman Antiquities*, 3.69.5, though they fail to mention the roof opening. Servius, *ad Aen.* 9.448, writes: "From the place where a portion of the Capitoline temple's forward-leaning roof looks towards Terminus's stone, *Unde in Capitolio prona pars tecti patet, quae lapidem ipsum Termini spectat.*"

[6] Dionysius, *Roman Antiquities*, 3.69; Livy, *ab Urbe Condita*, 1.38.

[7] For our 20 February 753 BCE—the year of Rome's foundation according to the tradition handed down by Varro—the Cosmos Programme has Arcturus rising at 18:34 with 45°.29 azimuth and the Sun setting at 17:45.
The azimuth is the angle between geographical north and the observed horizon point, moving clockwise from the North.

[8] Aveni-Romano 1994, p. 65.

[9] See, for all of these, Pliny, *Naturalis Historia*, 2.143.

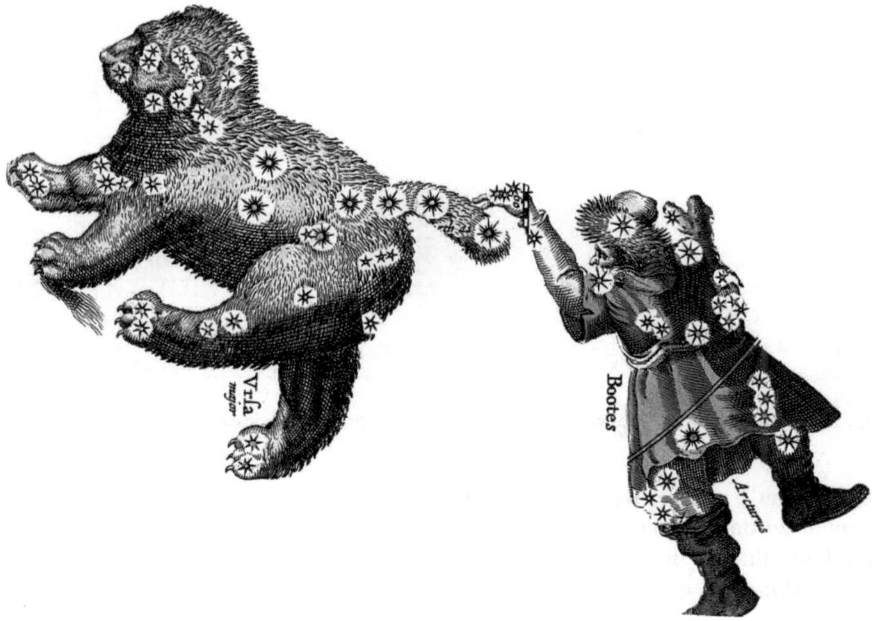

Fig. 25.1 *Arcturus*, the "guardian of the (Great) Bear", and *Ursa Major*—from Andrea Cellarius, *Atlas Coelestis seu Harmonia Macrocosmica*, 1660

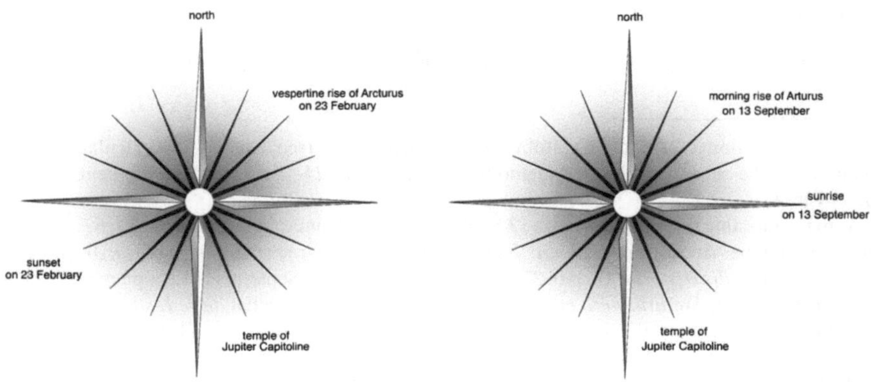

Fig. 25.2 The *disciplina etrusca*, the division of heavenly space, the vespertine rising of *Arcturus* on 23 February and the morning rise on 13 September. The *circle* of the horizon is divided into 16 segments, which are counted clockwise from the north. In Rome, the temple of Jupiter Capitoline opened in the direction that is the boundary for the seventh and eighth segments; around the eighth century BCE, the vespertine rising of Arcturus was in the direction that separated the second and third segments and the morning rising was in the direction that separated the 14 and 15th segments. (from *Les Etrusques* 1992; revised by the author)

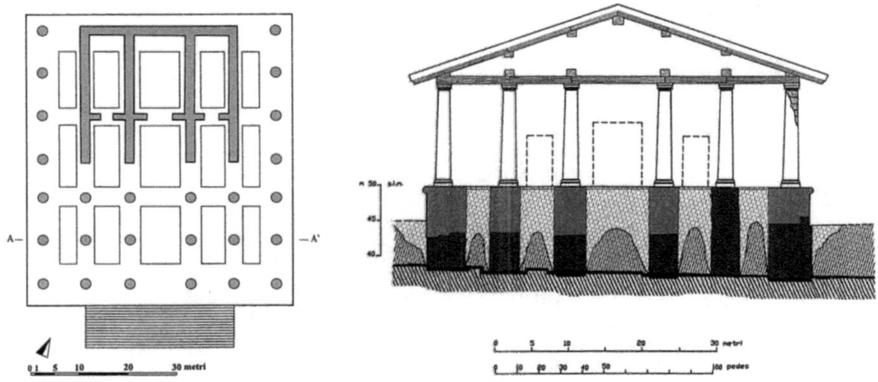

Fig. 25.3 The temple of Jupiter Capitoline: plan and hypothetical elevation. (from Cifani 2008)

And that's not all: we just saw that vespertine rise of Arcturus occurs for the *Terminalia* on February 23 when the Sun enter into the sign of Pisces. Now we could add that the morning rise of the same Arcturus occurs for the *Epulum Iovi Iunoni Minervae*, on 13 September (see Chap. 3), when the Sun enters the sign of Libra.[10]

The relationship between the celestial body, which undertakes its evening rising on 23 February and its morning rise on 13 September, and the Capitoline temple, which remains the residence of the old God *Terminus* celebrated at the *Terminalia* and becomes the residence of the new God *Iovis* celebrated with the *Epulum*, predates by a few centuries Ptolemy's assertion that

> the luminous and reddish star known as Arcturus is akin to Mars and Jupiter.[11]

---

[10] The celebration of Jupiter Capitoline on 13 September (see Chap. 15)—calculated, once again, using the second of the two methods listed in Chap. 24—occurs at the moment the Sun enters Libra.

As to the morning rise of Arcturus on 13 September, see Johannes Lydus, *de ostentis*, 67: "On the Ides of September Arcturus rises, *ειδοις Σεπτεμβριαις [...] δια της επιτολης του Αρκτουρου*. See also Pliny, *Naturalis Historia*, 18.310: "The day before the Ides of September [12 September] half of Arcturus rises [...], *Arcturus vero medius prid. Id [...]*"; and Columella, *de re rustica*, 11.2: "The fifteenth day before the Calends of October [17 September] Arcturus rises, *XV Kal. Oct. Arcturus exoritur.*"

On the topic of the morning rise of Arcturus, Pliny, *Naturalis Historia*, 18.311, adds: "The signal for the rise of this heavenly body will be given by observation of the departure of the swallows, given that they die if they are taken by surprise, *Signum orientis eius sideris servetur hirundinum abitus; namque deprehensae intereunt.*"

It follows that on 23 February in Rome the swallows arrive along with the vespertine rising of Arcturus, and on 13 September they depart with the morning rise.

Bickerman 1980, p. 114, provides the following data regarding Arcturus for the latitude of Rome (42°N) and for 500 BCE: morning rise ca. 16 Settembre, vespertine rise ca. 23 February.

[11] Ptolemy, *Tetrabiblos,* 1.9.15: [...] *ο δε λαμπρος και υποκιρρος καλουμενος Αρκτουρος τω τε του Αρεως και τω του Διος.*

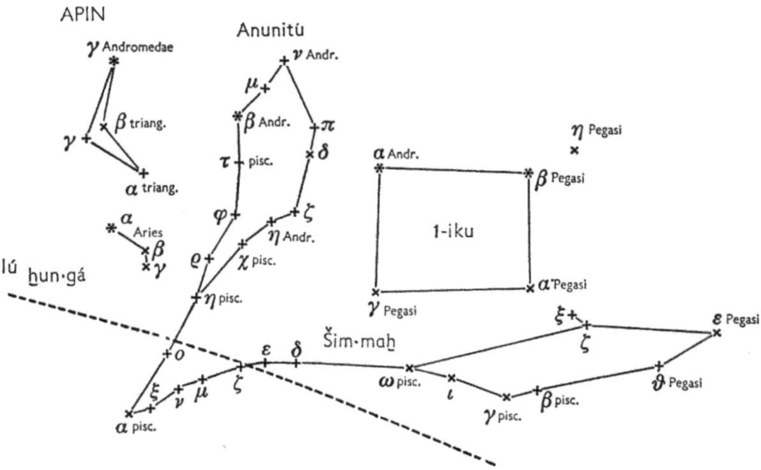

Fig. 25.4 The constellation of the "great swallow" in Mesopotamian astronomy. In Babylonia, the Southwestern branch of our constellation *Pisces*, together with the stars ε ζ θ from *Pegasus*, was known as *sim.mah*, the "great swallow", for its resemblance to a swallow; the Northwestern branch, plus a portion of *Andromeda*'s stars, was known as *A-nu-ni-tum*. (from Van der Waerden 1949)

Pisces, the sign that the Sun enters for the *Terminalia* on February 23, just as the swallows arrive in Rome, once again has a link to Babylon. The southwestern branch of the double constellation of Pisces, along with stars ε ζ θ from Pegasus, were known as *sim.mah*, or "big swallow" (Fig. 25.4), while the northwestern branch, along with a portion of the stars that make up Andromeda, were known as *A-nu-ni-tum*. Another, later, name for the constellation of Pisces, *Zib*, seems to connote the border for the series of twelve signs as "Boundary",[12] a definition close to Latin *Terminus*.

[12] The name of the southwestern branch of Pisces is in Van der Waerden 1949, pp. 6–26. The name *Zib* given to the constellation of Pisces in late Greek-Babylonian astronomy may be found in Allen 1963, p. 337; though this book has at least in part been superseded, this piece of information should nevertheless remain valid.

Chapter 26
The Vespertine Rising of Arcturus for the *Terminalia* on February 23 and the Relationship between the Romulean and Numan Calendars

The God *Terminus*'s unwillingness to vacate the position he had been allocated in a forgotten age, and the "small hole" opened in the ceiling of the new Capitoline temple so "that he may see naught above him but the stars", hark back to the earliest observations of celestial bodies in Rome. The vespertine rising of Arcturus for the *Terminalia* on February 23 allows us to identify a link between the Numan calendar and its predecessor, which is traditionally ascribed to the city's founder and first king, Romulus (753–716 BCE)

The Romulean year is described thus by Macrobius:

> There was a time when the Romans, thanks to Romulus, had their own 10 month year, beginning in March and lasting 304 days: 6 months—i.e. April, June, August, September, November and December—were 30 days long; 4 months—i.e. March, May, July and October—were 31 days long.[1]

From an astronomical standpoint, a 304 day year makes no sense at all: it is neither solar nor lunar, and it doesn't even last a whole number of lunations. Then there's the fact that 31 day months are not compliant with lunations, which last around 29.5 days; it would make far more sense for 30 day months alternating with 29 or 31 day months with 28 day months. In consequence, it is generally believed that the Romulean year was *not* astronomically-based.

And yet in another comment from Macrobius—one more important yet neglected comment, much like the rest of his writings on the Numan cycle (see Chap. 4)—we discover that a link does indeed exist between month and season in the Romulean year:

[1] Macrobius, *Saturnalia*, 1.12.3: *Romanos quoque olim auctore Romulo annum suum decem habuisse mensibus ordinatum: qui annus incipiebat a Martio et conficiebatur diebus trecentis quattuor, ut sex quidam menses, id est Aprilis Iunius Sextilis September November December, tricenum essent dierum, quattuor vero, Martius Maius Quintilis October, tricenis et singulis expedirentur.*
See also Censorinus, *de die natali*, 20.3.

L. Magini, *Stars, Myths and Rituals in Etruscan Rome*, Space and Society,
DOI 10.1007/978-3-319-07266-1_26

> Given that this number [304 days: author's note] agrees neither with the motion of the Sun nor with the rhythm of the Moon, at times it occurred that the cold part of the year took place in the summer months or, vice versa, the hot part of the year in the winter months.
>
> When this happened, a number of days as large as those necessary to return the season of the year to the particular climate of that month was allowed to be lost, without any monthly name.[2]

This comment leaves absolutely no room for doubt: every month in the Romulean calendar is associated with "the particular climate of that month", i.e. the appropriate type of weather, which means that *the Romulean year is indeed bound up with the motion of the Sun.*

In consequence, no 304 day year can be followed by a new 304 day year without a break—as is the case today in parts of the world where a purely lunar calendar is still in use. This is because without such a break, every month would slide back through the entire solar year, and could not therefore be associated with a specific "climate" or season: a year One lasting 10 months from March to December would be followed by a year Two in which March starts when January had started the previous year; in year Three, March would be where November had been two years previously, and so on. The obvious consequence is that it would have been impossible to associate a season with any given month in any stable form.

The final quote from Macrobius shows that the Romulean calendar covers 304 days of the solar year, divided up into 10 numbered months, which left 61 days "without any monthly name."

Something similar applied to a different civilization and tradition at the same time as the first Roman kings. In *Works and Days*, Hesiod writes:

> When Zeus [i.e. the Sun] has finished sixty
> wintry days after the solstice, then the star
> Arcturus leaves the holy stream of Ocean,
> and first rises brilliant at dusk;
> after him the shrilly wailing daughter of Pandion, the swallow,
> appears to men when spring is just beginning.[3]

The same occurs in Rome with the Numan calendar, in which the winter solstice falls on 21 December. As may be deduced from a reading of Varro's words on the topic:

> The time from the *bruma* until the Sun returns to the *bruma*, is called a year;

---

[2] Macrobius, *Saturnalia*, 1.12.39: *Sed cum is numerus neque solis cursui neque lunae rationibus conveniret, non nunquam usu veniebat ut frigus anni aestivis mensibus et contra calor hiemalibus proveniret; quod ubi contigisset, tantum dierum sine ullo mensis nomine patiebantur absumi quantum ad id anni tempus adduceret quo caeli habitus instanti mensi aptus inveniretur.*

[3] Hesiod, *Works and Days*, 564–9: *ευτ' αν δ' εξηκοντα μετα τροπας ηελιοιο / χειμερι' εκτελεσει Ζευς ηματα, δη ρα τοτ αστηρ / Αρκτουρος προλιπων ιερον ροον Ωκεανοιο / πρωτον παμφαινων επιτελλεται ακροκνεφαιος: / τον δε μετ ορθρογοη Πανδιονις ωρτο χελιδων / εσ φαος ανθρωποις, εαρος νεον ισταμενοιο.* Pandion, King of Attica, had his daughter turned into a swallow.

and of Ovid's:

> Midwinter is the beginning of the new Sun and the end of the old one;
> Phoebus [i.e. the Sun] and the year take their start from the same point;

and the Praenestine Calendar regarding the feast of *Divalia*, on 21 December:

> There are those who believe that the ceremony for this day is to celebrate the New year; it is evident, indeed, that [this day] is the start of the New year.[4]

In the Numan calendar, starting from the 21 December solstice, the Sun "finishes sixty wintry days" on 22 February,[5] the day that the swallows appear; the day after, Arcturus performs its vespertine rising. This is exactly as Pliny recounts it (see Chap. 25), and how Hesiod chronicled events too.[6] So, the period of time separating these two significant astronomical phenomena—the winter solstice and Arcturus's vespertine rising—corresponds fairly well to the 61 days missing from the Romulean calendar, "without any monthly name". The remaining 304 days, broken down into 10 numbered months, cover the rest of the year, from Arcturus's vespertine rising to the winter solstice.

It is here that the two calendars begin to display their fundamental differences, regardless of their chronological link and similar heritage.[7] The older of these years—the Romulean year—requires a re-anchoring every year through the observation of two significant astronomical phenomena if it is to comply with the movement of celestial bodies; however, Macrobius notes that this alone was not necessarily sufficient. The more modern of the two years—the Numan year—is far more similar to our own year, requiring solely the addition of intercalary days as prescribed by the rules in order to remain in sync with the motion of the Sun, Moon and planets (see Chap. 21) over a long cycle of years (see Chap. 4).

---

[4] Varro, *de lingua latina*, 6.8: *tempus a bruma ad brumam dum sol redit, vocatur annus. Bruma* is the "very short day", the shortest of the year; Varro also talks of *dies brumales*, the "shortest days", because of the difficulty in working out which one of them is the actual solstice.
Ovid, *Fasti*, 1.163–4: *Bruma novi prima est veterisque novissima solis; / principium capiunt Phoebus et annus idem.*
The Praenestine Calendar reads thus: S]UNT TAMEN, [QUI FIERI ID SACRU]M AIUNT OB AN[NUM NOVUM; MANI]FESTUM ESSE [ENIM PRINCIPIU]M [A]NNI NOV[I]. Mommsen's rather freighted addition on the basis of chronicles by Varro, Pliny (following on from Verrius Flaccus) and Macrobius, is universally accepted; indeed, as Warde Fowler 1908, p. 275, noted, "the Praenestine fragments clearly suggests the word '*annus*'."

[5] 9 days of December + 29 days of January + 22 days of February = 60 days.

[6] Even using modern-day calculations, at the time of Rome's first kings, 60 days elapsed between the winter solstice and the evening rising of Arcturus. For example, the day of the winter solstice—when the Sun sets at 4:46 p.m. with azimuth 238°.54—corresponds to our 22 December 754 BCE; the day of Arcturus's vespertine rising—at 6:34 p.m., 49 min after sunset—corresponds to our 20 February 753 BCE (see Chap. 25, no. 10). All data has been sourced from the Cosmos Programme.

[7] According to tradition handed down among others by Censorinus, *de die natali*, 20.4, the Romulean year was almost immediately amended by Romulus's successor, Numa Pompilius: "Afterwards, either by Numa [...] there was instituted a year of 12 months and 355 days, *Postea sive a Numa [...] XII facti sunt menses et dies CCCLV.*"

The vespertine rising of Arcturus, as the Sun moves into the sign of Pisces for the feast of *Terminalia*, marks the end of the Ancient Roman liturgical year. Terminus's inflexible and steadfast resolve not to cede his place even to Jupiter Optimus Maximus not only marks—in all likelihood—the point where the two calendars (Romulean and Numan) coincide, it shows the inalterable nature of the relationship between rite and the observation of celestial bodies. Ovid confirms this in his statement:

> From that time, Terminus, thou hast not been free to flit:
> abide in that station in which thou hast been placed.[8]

His is, however, a deceptive lack of motion that the slow, imperceptible and inexorable precession of the equinoxes undermines, year after year and century after century, as it alters the time and azimuth of the vespertine rising of all heavenly bodies, Arcturus in particular. With the passage of time, it is no longer possible to observe the first appearance of the celestial body in the East soon after sunset on 23 February from the "small hole" on the Capitoline temple roof (see again Fig. 25.3) above the altar to Terminus.

And yet, as Augustine noted almost a thousand years later,[9] *Terminus* would maintain his privilege of representing the "end of ephemeral things", while *Janus* represents the "beginning". The two gods remain united: *Terminus* is associated with a specific point in the heavenly vault, and a specific time of year; *Janus* is associated with the entire heavenly vault and its slow, ceaseless revolution for all time.

---

[8] Ovid, *Fasti*, 2.673–4: *Termine, post illud levitas tibi libera non est: / qua positus fueris in statione, mane.* "*Post illud*, from that time", regarding the god's refusal to move to make way for construction of the new temple.

[9] Augustine, *de Civitate Dei*, 7.7: *[…] ad eum* (scil. *Ianum) dicuntur rerum temporalium initia pertinere, fines vero ad alterum, quem Terminum vocant.*

Chapter 27
Reform of the Romulean Year and the Transition to the Numan Year

We began by stating that the Numan calendar is "the oldest Western calendar with a clear link to astronomy" (see Chap. 2). Now, on the contrary, we are obliged to acknowledge that the Romulean calendar also *had links* to astronomy.

The difference lies in the type of link:

- The Romulean calendar is based on the observation of two astronomical phenomena (the winter solstice and the vespertine rise of Arcturus)[1];
- The Numan calendar is based on knowledge regarding the different lengths of lunations and of the solar year, in addition to the motions of the other heavenly bodies.

In other words, the Romulean calendar is still a "primitive" calendar, even if the winter solstice and the rising of the brightest star in the northern hemisphere are astronomical "phenomena" of primary importance.[2] The Numan calendar, on the other hand, is not just a "modern" calendar, it is the direct progenitor of our modern-day calendar: the Gregorian calendar was begat by the Julian calendar, and the Julian calendar was begat by the Numan calendar.

In my opinion, this sudden and brusque transition from a "primitive" to a "modern" calendar is the biggest and most telling clue that the Etruscans hail from the Orient, and that they brought their cultural baggage to the people of Lazio who were soon to found the Eternal City. This topic, however, falls outside the scope of

---

[1] The winter solstice may be observed both in the East when the Sun rises, and in the West when it sets; the morning rise of Arcturus is observed by looking solely East for the star that rises just before the Sun. But the vespertine rise of Arcturus may be seen only by looking both West, as the sun sets, and East, where the star rises: the "small hole" Ovid mentions (see Chap. 25) must be on the highest point of the Jupiter Capitoline temple roof (Fig. 25.3).

[2] In its etymological sense, the Greek noun *phainomenon*, "that which appears, that which may be observed", in astronomy is more extensively interpreted as "that which may be observed and studied through direct observation."

L. Magini, *Stars, Myths and Rituals in Etruscan Rome*, Space and Society,
DOI 10.1007/978-3-319-07266-1_27

our research here, and requires an approach that specifically examines the origins of the Etruscan civilization and the foundation of Rome.[3]

The Numan calendar—as we have seen *ad abundantiam*—employs a system of intercalation based on a large quantity of refined knowledge about heavenly bodies. For this calendar, direct and "practical" ongoing observation is required solely to verify what we may call the abstract or "theoretical" results reached through well-known and well-codified calculations.

At this point, it becomes important to attempt to establish how Numa's "reform" made it possible to move from one calendar to the other. We will begin with Censorinus and Macrobius' writings,[4] which state that the Romulean year lasted 304 days; it began on 1 March and ended on 30 December, which was the last day of the year.

It is worth recalling the quotes from Varro, Ovid and the Praenestine Calendar that we saw in Chap. 26. Varro writes:

> The time from the *bruma* until the Sun returns to the *bruma*, is called a year;

Ovid writes:

> Midwinter is the beginning of the new Sun and the end of the old one;
> Phoebus [i.e. the Sun] and the year take their start from the same point;

and the Praenestine Calendar on 21 December writes:

> there are those who believe that the ceremony for this day is to celebrate the New Year; it is evident, indeed, that [this day] is the start of the New Year.

The only legitimate conclusion we may draw from these writings is that the winter solstice marked the end of the year—and therefore fell on 30 December—*in the Romulean calendar*. It may be objected that in actual fact, neither Varro nor Ovid specify exactly which "year" they are referring to. However, we may be certain about one thing: it is not the Numan year, in which the winter solstice

---

[3] See Magini 2011.

[4] Censorinus, *de die natali*, 20.2-4 recalls that the ten-month long Romulean year was "like the year of the Albans, from which the Roman year was descended, *ut tunc Albanis erat, unde orti Romani*". And Festus (150), representing Verrius Flaccus, says: "The month of March marked the start of the year in Latium, and did so after the foundation of Rome, *Martius mensis initium fuit anni et in Latio et post Romam conditam.*" See Warde-Fowler 1908, p. 5, no. 1.

As we saw, Censorinus goes on to explain: "These 10 months lasted 304 days, distributed thusly: March 31, April 30, May 31, June 30, Quintile [i.e. July] 31, Sextile [i.e. August] and September 30 each, October 31, and November and December 30. The four longer months were known as whole months; the other 6 months were known as incomplete months. Afterwards, Numa [...] created 12 months and 355 days [...], *Hi decem menses dies CCCIIII hoc modo habebant: Martius XXXI, Aprilis XXX, Maius XXXI, Iunius XXX, Quintilis XXXI, Sextilis et September tricenos, October XXXI, November et December XXX; quorum quattuor maiores pleni, ceteri sex cavi vocabantur. Postea sive a Numa [...] XII facti sunt menses et dies CCCLV [...]*"

Macrobius, *Saturnalia* 1.12.3 writes: "But Numa, his successor [...] added [...], *Sed secutus Numa [...] addidit [...]*" before continuing as Censorinus does.

occurs on 21 December. Once we have ruled out the Numan year, the only other year it could be in Rome is the old Romulean year.

Plutarch would appear to provide the casting vote for this thesis:

> But consider whether Numa may not have adopted as the beginning of the year that which conforms to our conception of the natural beginning.
>
> Speaking generally, to be sure, there is not naturally either a last or a first in a cycle; and it is by custom that some adopt one beginning of this period and others another.
>
> They do best, however, who adopt the beginning after the winter solstice, when the sun has ceased to advance, and turns about and retraces his course toward us.[5]

It could be posited that Plutarch is not talking about the end of the old year and the beginning of the new, but about the changes Numa made to the order of the months, whereby January and February preceded Romulus's first month (March). This objection, however, does not hold water, as a number of writings demonstrate that this order was not Numa's doing, but came into effect at a later date. In Numa's time, January and February followed December, and February was the last month of the year, as Varro testifies (see Chap. 7).

This leaves us with only one likelihood: Plutarch is providing us with that proof that we were looking for all along, namely, that Numa established the beginning of the year "after the winter solstice".

Let's look a little more closely at this. Let's imagine that Romulus chose the winter solstice as the last day of his year (30 December), and that Numa in his calendar reform, brought it forward to 21 December (under the Numan year).

To recap, according to tradition (see Chap. 2), Numa's reform consisted of the following:

- To begin with, Numa added 51 days to the 304 days of the Romulean year and obtained a 355 day lunar year;
- He then removed 1 day from the six 30 day months, recouping 6 days;
- He added these 6 days to the 51 already added to the Romulean year, giving $6 + 51 = 57$ days;
- Lastly, he split these 57 days into two new months, to create a 29 day January and a 28 day February.

Following this series of changes, the 304 day year divided into 10 months, plus the 60 days "without any monthly name", roughly made up a 355 day year, which was augmented by an average of 10.25 intercalated days. The Romulean calendar became the Numan calendar, and the Numan year applied intercalation on a 24 year cycle as we saw earlier (see Chap. 4).

---

[5] Plutarch, *Romanae Quaestiones*, 19: *Ορα δε μη μαλλον ο Νομας τη φυσει προσηκουσαν αρχην ελαβε του ετους ως προς ημας. καθολου μεν γαρ ουδεν εστι φυσει των εν κυκλω περιφ ερομενων ουτ εσατον ουτε πρωτον, νομω δ αλλην αλλοι του ρονου λαμβανουσιν αρην. αριστα δ οι την μετα τροπας χειμερινας λαμβανοντες, οπηνικα του προσω βαδιζειν πεπαυμενος ο ηλιος επιστρεφει και ανακαμπτει παλιν προς ημας.*

Table 27.1 Correspondences between the Romulean and Numan years

Romulean year		Numan year	
Interval	No. of days	Interval	No. of days
From 1 to 14 March	14	* * *	* * *
15 March >>>>>>>>>>>>> 1 March—Kalendae Marti			
From 15 March to 30 December	290	From 1 March to 21 December	290
30 December >>> winter solstice >>> 21 December—Divalia			
Days without a monthly name	60	Last 9 days of December + 29 days of January + 22 days of February	60
Vespertine rising of Arcturus >>> 23 February—Terminalia			
* * *	* * *	Last 5 days of February + 10,25 intercalary days	15,25
Total days	14 + 290 + 60 = 364	* * *	290 + 60 + 15,25 = 365,25

Table 27.1 summarizes the changes between the Romulean and the Numan years according to this thesis:

- In the Romulean year, the interval between 15 March and 30 December lasts 290 days; in the Numan year, the interval between 1 March and 21 December lasts 290 days[6];
- The Romulean year has 60 days (without a monthly name) between the winter solstice and the vespertine rising of Arcturus. As we saw in Chap. 26, the same number of days in the Numan year run from 21 December to 22 February, the day that the swallows appear, and, the day after, Arcturus performs its vespertine rising; these 60 days consist of the last 9 days of December, the 29 days of January and the first 22 days of February;
- In the Romulean year, the first 14 days of March, added to the following 290 days, make up the 304 days of the 10 month year. In the Numan year, the last 5 days of February, plus the added 10.25 average intercalated days (see Chap. 4), make a total of 15.25 days.

All in all, the Romulean year lasted 364 fixed days, while the Numan year, on average, lasted 365.25 days.

Table 27.1 also shows correspondences between specific days in both of these years:

---

[6] In the Romulean year: 17 days in March + 273 days in April, May, June, July, August, September, October, November and December = 290 days. In the Numan year: 269 days in March, April, May, June, July, August, September, October and November + 21 days in December = 290 days.

- The winter solstice fell on 30 December in the Romulean year and on 21 December in the Numan year;
- The last of the 60 days "without a monthly name" corresponds to the vespertine rising of Arcturus in the Romulean year, and to 23 February in the Numan year;
- 15 March in the Romulean year corresponds to 1 March in the Numan year.

Chapter 28
(Inevitable) Conclusion: *Per Aspera Ad Astra*

Among the many thousands of astronomical documents held in the royal libraries of Mesopotamia, the so-called *Astrolabes*—veritable lists of the northern, equatorial and southern constellations, or "three constellations for every month of the year", as the tablet inscriptions state—held a very special position. The *Astrolabes* list "relative positions, rising and setting times, the value for agriculture and mythical meanings" for each of 36 stars.

The astrolabe traditionally known as *Astrolabe B*, drawn up in Assur during the reign of Tiglath-Pileser (around 1,100 BCE) is closely linked to a chronicle written many centuries later by Diodorus Siculus. This has been observed and commented upon as follows:

> Astrolabe B, after having mentioned the stars of Ea, Anu, and Enlil, and having connected them to 12 months of the year, gives 36 statements on simultaneous risings and settings, of the following kind: 1 *iku* rises and *nin.mah* disappears; *dili.pat* rises and *zibanitum* disappears; APIN rises and *en.te.na bar.guz* disappears, etc. The stars thus linked together always stand just 6 months apart in the astrolabe scheme. This means that every star is supposed to be visible during 6 months and invisible during 6 months. Now this is sheer mythology. Zodiacal stars are visible 10 or 11 months, and more northern stars even longer. Also for daily risings and settings the statements in the text are nonsense.
>
> Myths have a longer life than astronomical theories, and so we must not be astonished to find the same star myth, enriched and modified, in Diodorus (2.30-1):
>
> Under the course on which these planets move, according to them, thirty stars, which they (the Chaldeans) designate as *Counselling Gods*; of these one half oversee the regions above the earth and the other half those beneath the earth, having under their purview the affairs of mankind and likewise those of the heavens; and every 10 days one of the stars above is sent as a messenger, so to speak, to the stars below, and again in like manner one of the stars below the earth to those above, and this movement of theirs is fixed and determined by means of an orbit which is unchanging forever. Twelve of these gods, they say, hold chief authority, and to each of these the Chaldeans assign a month and one of the signs of the Zodiac [...][1].

---

[1] This is excerpted from Van der Waerden 1949, pp. 6–26.

L. Magini, *Stars, Myths and Rituals in Etruscan Rome*, Space and Society,
DOI 10.1007/978-3-319-07266-1_28

Unfortunately, when it comes to the Etruscans, the chronicles are always later and indirect. Though Arnobius's writings were even later, there is no doubt that he has much of interest to say on the topic:

> In his 16th book, following Etruscan teachings, he [Nigidius Figulus; author's note] shows that there are four kinds of Penates; and that one of these pertains to Jupiter, another to Neptune, the third to the shades below, the fourth to mortal men, and then makes some unintelligible assertion […].
> Varro believes that they are the gods of whom we speak who are within, and in the innermost recesses of heaven, and that neither their number nor names are known. The Etruscans say that these are the *Consentes*, 'They who make a harmonious decision', and the *Complices*, 'They who are clutched together', naming them because they rise and fall together, six of them being male, and as many female, with unknown names and pitiless dispositions, but they are considered the counsellors and princes of Jove Supreme.[2]

We find that ever so slowly, the inexorable passage of time warps a rich, varied, harmonious and gratifying vision of the world. The loss of understanding, comprehension and recollection, along with the transition from one culture to another, impoverishes it to the point that it almost completely loses its extraordinary original value. What remained overcame the barrier of centuries but in so doing become almost unrecognizable and incomprehensible, retaining scant traces of what was once a majestic and brilliant conception of the relationship between man and the cosmos.

And yet, on high:

> Iuno, Vesta, Minerva, Ceres, Diana, Venus, Mars,
> Mercurius, Iovis, Neptunus, Vulcanus, Apollo[3]

continue to show themselves and show the way. As Ovid so nicely puts it,

> Ah, happy souls, who first took thought to know these things
> and scale the heavenly mansions![4]

---

[2] Arnobius, *adversus nations*, 3.40.1 and 3: *Nigidius […] in libro sexto exponit et decimo disciplinas Etruscas sequens, genera esse Penatium quattuor et esse Iovis ex his alios, alios Neptuni, inferorum tertios, mortalium hominum quartos, inexplicabile nescio quid dicens […] Varro qui sunt introrsus atque in intimis penetralibus caeli deos esse censet quos loquimur nec eorum numerum nec nomina sciri. Hos* Consentes *et* Complices *Etrusci aiunt et nominant, quod una oriantur et occidant una, sex mares et totidem feminas, nominibus ignotis et miserationis parcissimae; sed eos summi Iovis consiliarios ac principes existimari.*

[3] These are the names of the *Consentes Dii* of which Ennius writes in hexameter: Juno, Minerva, Ceres and Diana are represented by the various aspects of the Moon; Neptune, Vulcan and Apollo by the Sun; Vesta by the Earth; Mercury, Venus, Mars and Jupiter by their respective planets. Saturn would appear to be missing from this list.

[4] Ovid, *Fasti*, 1.297-8: *Felices animae, quibus haec cognoscere primis / inque domus superas scandere cura fuit*!

Post Scriptum: A Number of Observations, with Hindsight

Astronomia Etrusco-Romana was first published in Italian in 2003, following *Astronomy and Calendar in Ancient Rome—The Eclipse Festivals* in 2001, and *Le Feste di Venere—Fertilità femminile e configurazioni astrali nel calendario di Roma antica* in 1996. In nigh on a decade of "crazy, desperate" study, as Giacomo Leopardi would have it, I have reconstructed a solid framework of the Roman calendar's astronomical underpinnings, especially the Numan calendar. *Stars, Myths and Rituals in Etruscan Rome* makes only minimal adjustments to this framework, as well as adding some interesting elements to the fray.

Over this time—and at long last—there has been a radical change in our understanding of man's relationship with the heavens during the time of Rome's early kings. The eighth century BCE calendars that have survived the centuries are no longer viewed as the basic calendars of an agricultural and pastoral society, lacking in any consideration for heavenly phenomena or the movements of heavenly bodies; calendar feast days are no longer considered simple anniversaries of natural events, such as storing away grain or lambing time. On the contrary, the Romulean calendar demonstrates an awareness of a number of significant celestial phenomena, while the Numan calendar and cycle are a highly advanced—indeed, close to perfect—mechanism for monitoring observable movements in the solar system.

The end result of this research is a demonstration not just that Romans in Augustus' day were mistaken in their belief—asserted time and time again by Ovid, our best witness[1]—that Romans were uninterested in and had no understanding of astronomy. More importantly still, the end result is an acknowledgement that extraordinary attention was lavished on the heavens, and a specific body of knowledge accrued in the heart of the Mediterranean during ancient times.

Ultimately, Etrusco-Roman astronomy has been added to the canon of astronomical traditions, despite the fact that until recently nobody had so much as mentioned it; indeed, nobody had even suspected its existence. Etrusco-Roman astronomy has now taken its rightful place alongside astronomical approaches

[1] See Ovid, *Fasti*, 1.27: "To be sure, Romulus, thou wert better versed in swords than stars [...], *Scilicet arma magis quam sidera, Romule, noras [...]*"; and *Fasti*, 3.111-2: "The stars ran their courses free and unmarked throughout the year [...], *libera currebant et inobservata per annum / sidera [...]*"

L. Magini, *Stars, Myths and Rituals in Etruscan Rome*, Space and Society,
DOI 10.1007/978-3-319-07266-1

from around the world (the Dogon, Polynesians, American Indians, Maya, and Incas); ancient astronomical bodies of knowledge from the major Eastern civilizations (China and India); and astronomical bodies of knowledge from the classical Mediterranean (Egypt, Babylon, and Greece).

In this diverse panorama of astronomical bodies of knowledge, astronomy of Etruscan Rome possesses one unique characteristic: it is completely written in to its calendar. Legendary events that took place in Rome, almost all of which then shifted to the twilight zone between legend and history, were reiterated in the science of heavenly bodies—a science that did not have its own independent existence, but rather overlapped with and was ultimately subsumed by the perpetual repetition of fixed cadences in the *Feriale antiquissimum*. Equally unique, all of this took place at a time and in a location where writing did not yet exist. Calendar codification and a lack of writing are the two convergent themes that indicate the foreign provenance of this astronomical knowledge, whose traces were lost far before Ovid's day. Now, twenty-five centuries on, this knowledge is being rediscovered from the ephemeral and still mysterious life of the Numan calendar and its cycles.

Questions need to be answered about why this mysterious life was so brief, and why this knowledge disappeared and was utterly eclipsed; we should also look into why it has taken quite so long to rediscover this knowledge. The first of these questions has already been investigated,[2] and we return to these issues here in summary, recapping from the very beginning, that is to say, the provenance of the Numan calendar that the Etruscans brought with them from the East, and then passed on to the Romans.

The provenance can only be Babylonia; a comparison between the rituals of the "substitute of the king" and *Regifugium* and *Vestalia* provides the best evidence of this.[3] Babylonia is where astronomy and astrology were born, and where they developed earlier and more completely than in any other culture known today. It was from Babylonia that Pythagoras—Numa's real or presumed advisor—and other great pioneers of Greek and Western science drew so much of their wisdom. It was from Babylonia, directly or indirectly, that the "barbarian better than Pythagoras" came; if we are to believe Plutarch,[4] this barbarian may have tutored Numa in the two divine sciences in a Rome that was only just emerging from the darkness of prehistory. As sciences, they were and remained foreign, before becoming a patrimony restricted to an *élite* of initiates who held this knowledge and, with it, the power of "connecting the life of man to the life of the heavens". The *pontifices*—still today, albeit in another guise—continue to do this, as they did then, by transmitting their precious knowledge from mouth to mouth in a tight circle of those privy to sacred matters, and to the most sacred of all: the calendar.

The same applies to India, where in the *Jyotisavedanga* it is explained:

---

[2] Magini 2001, pp. 109–112.

[3] See Chap. 16.

[4] Plutarch, *Life of Numa* 1.

> The Veda have been revealed for the purpose of completing the sacrificial rites and these rites have been placed according to a temporal order. Thus, whoever is versed in astronomy, which is the science of the computation of time, knows the sacrifices.[5]

It applied in Rome as well, where the *pontifex minor*—the youngest and least experienced—was responsible for observing the first lunar crescent marking the kalends. It is not known what observations, registrations, computations and predictions were the preserve of the *pontifex maximus*. In any case, the deep meaning of Cicero's words is clear:

> Once few knew whether it was permitted or not to proceed according to the law; in reality, public festivals were not public. Those who were responsible for them held great power, and they were asked even the day, as was asked of the Chaldeans.[6]

The terse *tamquam a Chaldeis* is worth more than a hundred treatises on history, archeology and linguistics put together.

We should not forget that climate influences such matters: the climate in Italy and Rome is not the same as the climate in Mesopotamia or Egypt. The consequence of this it that it was no longer practically possible to keep the extended, ramified, and perhaps even obsessive activity of observation up endlessly, for a thousand and one nights and a thousand and one years; an activity which had made every city in the Near East a focal point for the observation, measurement, recording and prediction of celestial events. The science declined until it fell into obscurity, and the scientists dwindled in number until they disappeared. The pontifices were followed by the decemvirs, who were barely able to keep alive a simulacrum of a calendar, chock full of errors and omissions, and characterized as full of uncertainties and approximations by the historians of Rome.

The coup de grace finally came in 304 BCE with the publication of a new calendar, which was greatly desired and needed by the lower classes of merchants and peasants. After its announcement in its roughest and most simplified form—how else could it be made public?—the calendar was no longer subject to the corrections and adjustments it would periodically need, either by adding or subtracting days. Even if anybody had retained the skills necessary to work on the delicate and complex mechanism of the calendar, he would no longer have been allowed to do so because the political and economic interests at stake were so huge. The opinion and caprice of those in charge of managing the calendar of which Censorinus[7] writes represent but one part of the truth; the rest is, without doubt, the effective loss of the knowledge needed to maintain the calendar in working order.

---

[5] Cited from Bezza, *Introduction* to al-Biruni, *Gli astri, il tempo, il mondo*, p. 8.

[6] Cicero, *Pro Murena*, 11.25: *posset agi lege necne pauci quondam sciebant; fastos enim vulgo non habebant. erant in magna potentia qui consulebantur, a quibus etiam dies tamquam a Chaldeis petebatur.*

[7] Censorinus, *de die natali*, 20.7.

So much for the "brief and happy life" and subsequent unravelling of the relationship between astronomy and the Numan calendar. As for its rediscovery, one may legitimately wonder why it took quite so long, especially compared to how long ago similar scientific discoveries were made in the 20th century: more than 100 years have elapsed since Strassmeier, Epping and Kugler's discovery of Babylonian astronomy; over 70 years since Dogon and Polynesian astronomy were discovered; 50 years since Mayan astronomy was discovered, and so on.

Indeed, there have been times between 304 BCE (which we may assume to be the date of the disappearance) and 1996 (when this rediscovery was first mooted) when it may have occurred but didn't: for example, in Julius Caesar's time, when a decision was made to reform the calendar; between the third and fourth centuries CE, when the Empire was attempting to stave off decline and ultimate collapse; after the Italian Renaissance, when classical heritage as a whole underwent reassessment; in the 19th century, when modern classical studies truly began; or in the latter half of the 20th century, after "hypercriticism" had run its course, and new, well-founded faith was vested in "sources", that is to say, tangible evidence from classical times.

Why did this rediscovery fail to take place earlier? Because in Caesar's day, the concern was to correct crude errors that had accumulated over time and led to an unacceptable gap between the theoretical calendar date and the actual time of the year, in an attempt to salvage fragments of knowledge which had long since been lost in proto-history. Between the third and fourth centuries CE, the primary focus was on retrieving as much information as possible from the past, rather than reprocessing and inserting such information into a worldview that had been lost forever. In the Renaissance, it was the more philosophical and aesthetic aspects of classical thought that appealed as part of the radical revision of rigid thought processes bequeathed by the dark ages. In the 19th century, the dominant idea was to accumulate and select elements that had survived the years, sift through them—in some cases by comparing them with approaches evident in contemporary "primitive" cultures—and then reorder and catalogue them, at last, in accordance with what were considered rational and modern criteria. Last but not least, in the latter part of the 20th century, a renewed focus on text led to a more meticulous, richer and more complete understanding of the many sources available, while at the same time favouring slower and closer individual examination as opposed to a global overview.

In order of importance, the reasons for this long period of oblivion and the failure to rediscover this knowledge at each of these times may be ascribed to: (1) a prejudicial belief in Roman astronomical ignorance; (2) an equally prejudicial belief that each branch of knowledge—physics, mathematics, nature-related or philosophical—came to the Romans from the Greeks; (3) the even more prejudicial belief that the Greeks invented an entire worldview *ex novo*—a belief that was understandably ours—leading to the inevitable reaffirmation of an indissoluble link between Aristotle and the Fathers of the Church.

We should not ignore other fairy tales too: that the Romans possessed a tangible, earthy view of the world because they were peasants, shepherds, warriors and lawmakers, in other words, because they were simply uninterested in abstractions or theoretical constructs; or the total unreliability of written sources, owing to their relative lateness, their isolation, their confusion and contradictory nature.

Having cast these prejudices aside, the certainty of the secret "meaning" of things and—in the case of astronomy and the calendar—the strength of the numbers themselves would have been more than sufficient, as indeed they proved to be, to dispell the mystery—a mystery that many claimed did not even exist.

The fact of the matter is that courage and imagination—freedom of thought—was required to replace "whys" based on negativity and impediment with more positive and enquiring "whys". Why is the goddess *Fortuna* accompanied by *Necessitas* and *Spes*? Why did *Fortuna*, a foreigner to Rome, have thirteen feast days in her honour in the Numan calendar? Why is she associated with other female divinities such as *Venere Verticordia* and *Mater Matuta* on two of these feast days? Why are these feast days separated by a 72-day interval? And, most importantly, why are certain feast days named as they are: *Veneralia*, *Matralia*, *Matronalia* and *Liberalia*?

In conclusion, far better to cast aside ideological bonds and fly free and unencumbered in the aerial byways of research.

L.M.—Torrimpietra, 15 March 2014

The following table is divided into 12 sections, one for each month of the *Numan* year. The four columns present the following data:

- The day of the year and the nundinal letter (which continues from one month to the next, but always starts once more from the letter A in the new year). It should be noted that every year has its own nundinal letter, so that the first market day each year may duly take place eight days after the last market of the preceding year, and then continue every eight days;
- The day of the month;
- The feast day and/or divinity celebrated (*Feriale antiquissimum* feast days are in bold);
- Astronomical phenomena and public or private occurrences:
- Solar (◯) and lunar (●) eclipses are stated by the year in the Numan cycle in which they take place, and a progressive number. As elsewhere in this book, solar eclipses are annotated in Roman type, lunar eclipses in italics, total eclipses in bold, and partial eclipses in light type;
- Planetary phenomena for VenusVenus (planet) take place in Year Zero, Minus One and Plus One (indicated as Year Zero, Year −1 and Year +1); the placement of these events is yet to be identified within the Numan cycle

Summary Chart The months of the Numan year, with feast days, astronomical phenomena and public and private occurrence

March			
Day of the year + nundinal letter	Day of the month	Festivity or divinity celebrated	Astronomical phenomenon and public or private events
1 A	1	FERIAE MARTI Matronalia Ancilia Ignis Vestae	Year 01: Start of the Saros Cycle and of the revolution of the lines of nodes and apsides **1st**○ with the Sun at the descendant node (?) New Moon at the node; node close to the Gamma Point; Moon at the major standstill Year 47: Mercury returns to its initial position End of the 9° lunation of pregnancy , feast of preparation for birth
2 B	2		
3 C	3		
4 D	4		Year 80: Mars returns to its initial position
5 E	5		
6 F	6	Supplicatio Vestae	
7 G	7	Vediovi	
8 H	8		Years 09 and 10: *17° and* ***19°*** ●
9 A	9	Ancilia moventur	Start of the solar year/Birth festival
10 B	10		
11 C	11		
12 D	12		
13 E	13		
14 F	14	EQUIRRIA	
15 G	15	FERIAE IOVI Annae Perennae	Year 19: **49th/1st**○, start of the new Saros Cycle Sun in Ari 0° End of female initiation ceremonies
16 H	16	ad Argeos	Years 17 and 18: 43rd and **47th**○
17 A	17	LIBERALIA	*Dies lustricus* for men End of male initiation ceremonies
18 B	18		
19 C	19	QUINQUATRUS	
20 D	20		
21 E	21		

(continued)

Summary Chart (continued)

March			
Day of the year + nundinal letter	Day of the month	Festivity or divinity celebrated	Astronomical phenomenon and public or private events
22 F	22		Year 10: 26th ○
23 G	23	TUBILUSTRIUM	Year 09: **22nd**○
24 H	24	Q.R.C.F.	Spring equinox (Columella)
25 A	25		Spring equinox (Columella, Pliny)
26 B	26		Spring equinox (Ovid)
27 C	27		
28 D	28		
29 E	29		
30 F	30	Iani	
31 G	31	Lunae	Years 17 and 18: ***34th*** and *36th* ●

(continued)

Summary Chart (continued)

April			
Day of the year	Day of the month	Festivity or divinity celebrated	Astronomical phenomenon and public or private events
32 H	1	Veneralia/Veneri Verticordiae Fortunae Virili	Year Zero: Venus/Hesperus at lower conjunction Feast of matrimony
33 A	2		
34 B	3		
35 C	4	Start of Ludi Matri Magnae	
36 D	5	Fortunae Publicae	
37 E	6		
38 F	7		
39 G	8		
40 H	9		
41 A	10	End of Ludi Matri Magnae	
42 B	11		
43 C	12	12–19. Ludi Cereales	
44 D	13	Iovi Victori Iovi Libertati	
45 E	14		
46 F	15	FORDICIDIA	Six lunations before and after the *October Equus* Year 17: 44th ○
47 G	16		Years 07 and 08: **18th and 20th**○ with the Sun at the ascendant node (?)
48 H	17		
49 A	18		
50 B	19	CERIALIA/ Libero Liberae	
51 C	20		
52 D	21	PARILIA Natalis Urbis	½ an eclipse year before *October Equus*
53 E	22		
54 F	23	VINALIA/Veneri Feriae Iovi	

(continued)

Summary Chart (continued)

April			
Day of the year	Day of the month	Festivity or divinity celebrated	Astronomical phenomenon and public or private events
55 G	24	Pueri lenonii	Year 16: *32nd* ●
56 H	25	ROBIGALIA	Sun in Tau 10° (Pliny)
57 A	26		
58 B	27		
59 C	28	Start of Ludi Florae	
60 D	29		

(continued)

Summary Chart (continued)

May			
Day of the year	Day of the month	Festivity or divinity celebrated	Astronomical phenomenon and public or private events
61 E	1	Laribus Praestites Bonae Deae	Year 08: *15th* ●
62 F	2		Year 07: ***13***th ●
63 G	3	End of Ludi Florae	
64 H	4		
65 A	5		
66 B	6		
67 C	7		
68 D	8		
69 E	9	LEMURIA	
70 F	10		Years 06, 15 and 16°:14th, **39**th and **41st**○
71 G	11	LEMURIA	
72 H	12		
73 A	13	LEMURIA	
74 B	14	Argeis	
75 C	15	Feriae Iovi Mercury Maiae	
76 D	16		
77 E	17		
78 F	18		
79 G	19		
80 H	20		
81 A	21	AGONALIA Vediovi	
82 B	22		
83 C	23	TUBILUSTRIUM Volcano	
84 D	24	Q.R.C.F.	
85 E	25	Fortunae Publicae	Years 05, 06 and 15:*9th*, ***11th*** and *30th* ●
86 F	26		
87 G	27		
88 H	28		
89 A	29	Ambarvalia	
90 B	30		
91 C	31		

(continued)

Summary Chart (continued)

June			
Day of the year	Day of the month	Festivity or divinity celebrated	Astronomical phenomenon and public or private events
92 D	1	Kal. Fabariae/Carna Marti Iunoni Monetae	Year −1: Venus at higher conjunction
93 E	2		Year 14: **37th**○
94 F	3	Bellonae	Year 13: 33rd ○
95 G	4	Herculi Mag. Custodi	
96 H	5	Dio Fidio	
97 A	6		
98 B	7	Vesta aperitur Ludi piscatorii	
99 C	8	Menti	Year 06: 15th ○
100 D	9	VESTALIA	Year 05: **12th**○
101 E	10		
102 F	11	MATRALIA Matri Matutae and Fortunae	Year Zero: Venus at greatest western elongation Feast of conception
103 G	12		
104 H	13	IOVI EPULONI Quinquatrus minusc.	
105 A	14		
106 B	15	Q.ST.D.F. Vesta clauditur	End of *sar puhi*
107 C	16		
108 D	17		Years 13 and 14: ***26th*** and ***28th*** ●
109 E	18	Annae sacrum	
110 F	19	Minervae	
111 G	20	Summano	Year 10: node in Ari 0° and new Moon at the major standstill
112 H	21		
113 A	22		
114 B	23		
115 C	24	Fortis Fortunae	Summer solstice (Columella, Pliny)
116 D	25		Summer solstice (Columella)
117 E	26		Summer solstice (Columella, Ovid)
118 F	27	Laribus Viae Sacrae Iovi Statori	
119 G	28		
120 H	29		

(continued)

Summary Chart (continued)

July			
Day of the year	Day of the month	Festivity or divinity celebrated	Astronomical phenomenon and public or private events
121 A	1	Iunoni Felicitati	
122 B	2		
123 C	3		Year 13: 34th ○
124 D	4		Years 3 and 4: **8th and 10th**○
125 E	5	POPLIFUGIA	
126 F	6	Start of Ludi Apollinares Fortunae Muliebri	
127 G	7	Nonae Capratinae Palibus duobus	
128 H	8	Vitulatio	
129 A	9		
130 B	10		
131 C	11		
132 D	12		Year 12: *24th* ●
133 E	13	End of Ludi Apollinares	
134 F	14		
135 G	15	Equitum probation	
136 H	16		
137 A	17	Honori	
138 B	18	Dies Alliensis	Year 4: *7th* ●
139 C	19	LUCARIA	Year 3: ***5th*** ●
140 D	20		
141 E	21	LUCARIA	
142 F	22	Concordiae	
143 G	23	NEPTUNALIA	
144 H	24		
145 A	25	FURRINALIA	
146 B	26		
147 C	27		Years 2, 11 and 12: 4th, **29th** and **31st**○
148 D	28		
149 E	29		
150 F	30	Fortunae Huiusce Diei	
151 G	31		

(continued)

Summary Chart (continued)

August			
Day of the year	Day of the month	Festivity or divinity celebrated	Astronomical phenomenon and public or private events
152 H	1	Spei Victoribus duabus	
153 A	2		
154 B	3	Augurium canariae	
155 C	4		
156 D	5	Saluti	
157 E	6		
158 F	7		
159 G	8	Soli Indigeti	
160 H	9		
161 A	10	Arae Opis et Cereris	
162 B	11		Years 01, 02 and 11: *1st, 3rd and 22nd* ●
163 C	12	Herculi Invicto	
164 D	13	Dianae/Vortumno/Fort. Eq./ Herc.Vic./Cast. Poll./Camenis	
165 E	14		
166 F	15		
167 G	16		
168 H	17	PORTUNALIA	
169 A	18		
170 B	19	VINALIA/Veneri FERIAE IOVI	Years 09 and 10: 23rd and **27th**○
171 C	20		
172 D	21	CONSUALIA	
173 E	22		
174 F	23	VOLCANALIA Maiae/Opis Op./Hora Quir.	
175 G	24	Mundus patet	
176 H	25	OPICONSIVIA	
177 A	26		Year 01 and 02: **2nd** and 4th ○
178 B	27	VOLTURNALIA	
179 C	28	Solis et Lunae	
180 D	29		

(continued)

Summary Chart (continued)

September			
Day of the year	Day of the month	Festivity or divinity celebrated	Astronomical phenomenon and public or private events
181 E	1	Iovi Tonanti Iunoni Reginae Feriae Iovi	
182 F	2		
183 G	3		
184 H	4	Start of Ludi Romani	
185 A	5	Iovi Statori	Years 09 and 10: *18th and 20th*●
186 B	6		
187 C	7		
188 D	8		
189 E	9		
190 F	10		
191 G	11		
192 H	12		
193 A	13	Iovi, Iunoni, Minervae epulum End of Ludi Romani	Year 18: 48th ○ Year 18: End of the Saros cycle Sun in Lib 00° Morning rise of Arcturus
194 B	14	Equorum probation	Year 17: 45th ○
195 C	15		
196 D	16		
197 E	17		
198 F	18		
199 G	19		
200 H	20		Year 9: 24th ○
201 A	21		
202 B	22		
203 C	23	Marti/Neptunus Apollini/Latonae Felicitati Iovi Stat. Iunoni Reg.	
204 D	24		
205 E	25		
206 F	26		
207 G	27		
208 H	28		Years 17 and 18: *35th* and *37th* ●
209 A	29		Year 08: *16th* ●

(continued)

Summary Chart (continued)

October			
Day of the year	Day of the month	Festivity or divinity celebrated	Astronomical phenomenon and public or private events
210 B	1	Tigillo Sororio/Iuno Sororia Fidei	Start of male and female initiations
211 C	2		
212 D	3		
213 E	4	Ieiunium Cereris	
214 F	5	Mundus patet	
215 G	6		
216 H	7	Iovi Fulguri Iunoni Curiti	
217 A	8		
218 B	9	Genio Publ./Faust. Felicitati Veneri Victrici Apollini	
219 C	10	Iunoni Monetae	
220 D	11	MEDITRINALIA	
221 E	12		
222 F	13	FONTINALIA	
223 G	14	Penatibus	Year 17: 46th ○ Year 19: end of the revolution of the line of nodes
224 H	15	FERIAE IOVI October Equus	Years 07 and 08: 19th and 21st ○ Sun at the descendent node (?)
225 A	16		
226 B	17		
227 C	18		
228 D	19		ARMILUSTRIUM
229 E	20		
230 F	21		
231 G	22		
232 H	23		Year 16: *33rd* ●
233 A	24	Favori/Veneri Erucinae	Year 15: *31st* ●
234 B	25		
235 C	26		
236 D	27		
237 E	28		
238 F	29		
239 G	30		Year 07: *14th* ●
240 H	31		

(continued)

Summary Chart (continued)

November			
Day of the year	Day of the month	Festivity or divinity celebrated	Astronomical phenomenon and public or private events
241 A	1		
242 B	2		
243 C	3		
244 D	4	Start of Ludi Plebei	
245 E	5		
246 F	6		
247 G	7		Years 06, 15 and 16: 16th, 40th and 42nd ○
248 H	8	Mundus patet	
249 A	9		
250 B	10		
251 C	11		
252 D	12		
253 E	13	Iovi epulum Fortunae Primigeniae Feroniae/Pietati	
254 F	14	Equorum probation	
255 G	15		
256 H	16		
257 A	17	End of Ludi Plebei	
258 B	18		
259 C	19		
260 D	20		
261 E	21		
262 F	22		Years 05 and 06: *10th and 12th* ●
263 G	23		
264 H	24		
265 A	25		
266 B	26		
267 C	27		
268 D	28		
269 E	29		

(continued)

Summary Chart (continued)

December			
Day of the year	Day of the month	Festivity or divinity celebrated	Astronomical phenomenon and public or private events
270 F	1	Neptuno/Veniliae/ Salaciae Pietati Fortuna Muliebri	Year 14: 38th ○
271 G	2		Year 13: 35th ○
272 H	3	Sacra Bonae Deae	
273 A	4		
274 B	5	Faunalia rustica	
275 C	6		
276 D	7		
277 E	8	Tiberino/Gaiae	Years 5 and 6: 13th and 17th ○
278 F	9		
279 G	10		
280 H	11	AGONALIA INDIGETI	
281 A	12	Conso	
282 B	13	Telluri et Cereri	
283 C	14		
284 D	15	CONSUALIA	
285 E	16		Year 14: *29th* ●
286 F	17	SATURNALIA	Year 4 and 13: *8th* and *27th* ●
287 G	18		
288 H	19	OPALIA Iuventati	
289 A	20		
290 B	21	DIVALIA Angeronalia	Winter solstice and start of the New Year (*Fasti Praenestini*)
291 C	22	Laribus Permarinis	
292 D	23	LARENTALIA/ FERIAE IOVI Dianae/Iunoni Reginae Tempestatibus	
293 E	24		
294 F	25		
295 G	26		
296 H	27		
297 A	28		
298 B	29		

(continued)

Summary Chart (continued)

January			
Day of the year	Day of the month	Festivity or divinity celebrated	Astronomical phenomenon and public or private events
299 C	1	Aesculapio/Coronidi Vediovi	
300 D	2		Year 13: 36th ○
301 E	3	Start of Laralia or Ludi Compitales	Year 4: 11th ○
302 F	4		Year 3: 9th ○
303 G	5	End of Laralia or Ludi Compitales Vicae Potae	Year 09: End of the revolution of the line of apsides Moon at the apogee
304 H	6		
305 A	7	Iano Patri	
306 B	8		
307 C	9	AGONALIA/Iano	
308 D	10		
309 E	11	CARMENTALIA/ Carmenta Antevorta Iuturnae	Year +1: Venus at greatest western elongation Year 12: *25th* ●
310 F	12		Year 11: *23rd* ●
311 G	13	Iovi Statori	Feast of the seventh month of pregnancy
312 H	14		
313 A	15	CARMENTALIA/ Carmenta Postvorta	Year −1: Venus at greatest eastern elongation
314 B	16		
315 C	17		
316 D	18		Year 3: *6th* ●
317 E	19		
318 F	20		
319 G	21		
320 H	22		
321 A	23		
322 B	24		
323 C	25		
324 D	26		Years 02 and 12: 6th and 32nd ○
325 E	27	Castori et Polluci	Year 11: 30th ○
326 F	28		
327 G	29		

(continued)

Summary Chart (continued)

February			
Day of the year	Day of the month	Festivity or divinity celebrated	Astronomical phenomenon and public or private events
328 H	1	Iunoni Sospitae Matri Reg.	
329 A	2		
330 B	3		
331 C	4		
332 D	5	Concordiae	
333 E	6		
334 F	7		
335 G	8		
336 H	9		
337 A	10		
338 B	11		
339 C	12		Year 02: *4th* ●
340 D	13	Start of PARENTALIA Fauno et Iovi Mundus aperitur	Year 01°: *2nd* ●
341 E	14		
342 F	15	LUPERCALIA	
343 G	16		
344 H	17	QUIRINALIA Fornacalia or Stultorum feriae	
345 A	18		
346 B	19		
347 C	20		Year 10: 28th ○
348 D	21	End of PARENTALIA/ FERALIA Tacitae Mutae Matris Larum	Year 09: 25th ○
349 E	22	Caristia or Cara cognatio	
350 F	23	TERMINALIA	Sun in Pis 00° Evening rise of Arcturus
- - - -	- - -	- - - - - - - - - -	- - - - - - - - - -

(continued)

Summary Chart (continued)

February			
Day of the year	Day of the month	Festivity or divinity celebrated	Astronomical phenomenon and public or private events
22 or 23 intercalary days in years 2, 4, 6, 8, 10, 12, 14, 16, 19, 22 and 24			4 intercalary year 02: 7th ○ 12 intercalary year 10: *21st* ●
- - - -	- - -	- - - - - - - - - -	- - - - - - - - - -
351 G	24	REGIFUGIUM	Start of *sar puhi*
352 H	25		
353 A	26		
354 B	27	EQUIRRIA	Year 01: 3rd ○
355 C	28		Years 59, 83 and 24: Saturn, Jupiter and Venus return to their initial positions

Bibliography

Augustine, *de civitate dei* (from the Internet - translated by the author)
Arnobius, *adversus nationes* (from the Internet - translated by the author)
Catullus, *Carmina*, edited by G. Chiarini, Frassinelli, Milan
Censorinus, *de die natali*, edited by V. Fontanella, Zanichelli, Bologna
Cicero, *de divinatione*, edited by S. Timpanaro, Garzanti, Milan
Claudian, *de raptu Proserpinae*, edited by F. Serpa, Rizzoli, Milan
Columella, *de re rustica*, translation by R. Calzecchi Onesti, Einaudi, Turin
Dionysius, *Roman Antiquities(romaikes archaiologias)*, edited by E. Cary, Heinemann, London-Cambridge
Aulus Gellius, *Noctes Atticae*, edited by L. Rusca, Rizzoli, Milan
Geminus, *Introduction to the Phenomena (Eisagoge eis ta phainomena)*, edited by G. Aujac, Les Belles Lettres, Paris
Hesiod, *Works and Days (Erga kai Emera)*, edited by G. Arrighetti, Garzanti, Milan
Hesiod, *Theogonia*, edited by G. Arrighetti, Rizzoli, Milan
Joannes Lydus, *de mensibus*, in G. Bezza, 1995
Joannes Lydus, *de ostentis* (from the Internet - translated by the author)
Inniorfici (Orphichimnes edited by G. Ricciardelli, Fondazione Valla/Mondadori, Milan
Livy, *ab Urbe condita (Storia di Roma)*, edited by C. Moreschini, translation by M. Scandola, Rizzoli, Milan
Macrobius, *Saturnalia*, edited by N. Marinone, UTET, Turin
Manilius, *Astronomica*, edited by G. P. Goold, Heinemann, London-Cambridge
Manilius, *Astronomica*, edited by S. Feraboli, E. Flores, R. Scarcia, Fondazione Valla/ Mondadori, Milan
Martial, *Epigrammi*, Rizzoli, Milan
Ovid, *I Fasti*, trans. L. Canali, edited by M. Fucecchi, Rizzoli, Milan
Ovid, *Fasti*, edited by J. G. Frazer, Heinemann, London-Cambridge
Ovid, *Les Fastes*, edited by R. Schilling, Les Belles Lettres, Paris
Pliny, *Naturalis Historia*, edited by A. Aragosti, Einaudi, Turin
Plutarch, *Le vite di Teseo e di Romolo*, edited by C. Ampolo and M. Manfredini, Fondazione Valla/Mondadori, Milan
Plutarch, *Il volto della luna (de facie)*, edited by L. Lehnus, Adelphi, Milan
Plutarch, *Quaestiones Romanae (Moralia I)*, edited by F.C. Babbitt, Heinemann, London-Cambridge
Plutarch, *Questioni romane*, edited by N. Marinone, Rizzoli, Milan
Plutarch, *de Fortuna Romanorum (La Fortuna dei Romani)*, edited by G. Forni, D'Auria, Naples
Porphyry, *L'antro delle ninfe* (*On the Cave of the Nymphs)*, edited by L. Simonini, Adelphi, Milan

L. Magini, *Stars, Myths and Rituals in Etruscan Rome*, Space and Society,
DOI 10.1007/978-3-319-07266-1

I Presocratici, edited by A. Lami, Rizzoli, Milan
Ptolemy, *Tetrabiblos*, edited by S. Feraboli, Fondazione Valla/Mondadori, Milan
Seneca, *Questioni Naturali*, edited by D. Vottero, Tea, Milan
Varro, *de lingua latina*, edited by R. G. Kent, Heinemann, London-Cambridge
Virgil, *Bucolica*, translation by L. Canali, Rizzoli, Milan
Virgil, *Georgica*, translation by L. Canali, Rizzoli, Milan
Vitruvius, *De architectura*, translation by A. Corso and E. Romano, Einaudi, Turin
Allen **1963**: R.H. Allen, *Star Names - Their Lore and Meaning*, Dover New York
Ampolo **1988**: C. Ampolo, *Commento a* Le vite di Teseo e di Romolo *di Plutarco*, Fondazione Valla/Mondadori, Milan
Astronomia ***1963****: Astronomia*, edited by K. Stumpff, Feltrinelli, Milan
Aveni **1993**: A. Aveni, *Gli imperi del tempo*, Dedalo, Bari
Aveni **1997**: A. Aveni, *Astronomia nelle Americhe*, in C. Walker (edited by), *L'astronomia prima del telescopio*, Dedalo, Bari
Aveni **2000**: A. Aveni, *Scale fino alle stelle - La contemplazione del cielo in tre grandi civiltà antiche*, Corbaccio, Milan
Aveni-Romano **1994**: A.Aveni -G. Romano, *Orientazioni di templi e rituali etruschi*, "Rivista di Archeologia", XVIII
Bartholomae **1961**: C. Bartholomae, *Altiranisches Woerterbuch*, Berlin
Bezza **1995**: G. Bezza, *Arcana Mundi - antologia del pensiero astrologico antico*, Rizzoli, Milan
Bianchi Bandinelli **1970**: R. Bianchi Bandinelli, *Roma - La fine dell'arte antica*, Rizzoli, Milan
Bianchi Bandinelli -Giuliano **1985**: R. Bianchi Bandinelli-A. Giuliano, *Etruschi e Italici prima del dominio di Roma*, Rizzoli, Milan
Bianchi Bandinelli **1985a**: R. Bianchi Bandinelli, *Roma - L'arte romana nel centro del potere*, Rizzoli, Milan
Bickerman **1980**: E.J. Bickerman, *Chronology of the Ancient World*, Thames and Hudson, London
Bittel **1977**: K. Bittel, *Gli Ittiti*, Rizzoli, Milan
Blacker-Loewe **1978**: *Antiche Cosmologie*, (edited by) C. Blacker-M. Loewe, Astrolabio, Rome
Bottéro **1982**: J. Bottéro, *Sintomi, segni, scritture nell'antica Mesopotamia*, in *Divinazione e razionalità*, Einaudi, Turin
Bottéro **1991**: J. Bottéro, *Il sostituto reale e la sua sorte,* in *Mesopotamia - La scrittura, la mentalità e gli dei*, Einaudi, Turin
Brandenstein-Mayrhofer **1964**: W. Brandenstein-M. Mayrhofer, *Handbuch des Altpersischen*, Harrasowitz, Wiesbaden
Britton-Walker **1996**: J. Britton-C. Walker, *Astronomy and Astrology in Mesopotamia,* in C. Walker (edited by) *Astronomy Before the Telescope,* British Museum Press London
Buck **1988**: C.D. Buck, *A Dictionary of selected Synonyms in the principal Indo-European Languages*, Chicago-London
Caffarello **1975**: N. Caffarello *Avviamento allo studio della lingua etrusca*, Olschki, Florence
Carpenter **1996**: T. H. Carpenter, *Art and Myth in Ancient Greece*, Thames and Hudson, London
Castiglioni-Mariotti **1990**: L. Castiglioni-S. Mariotti, *Vocabolario della lingua latina*, Loescher, Turin
Cavalli Sforza **2000**: L.L. Cavalli Sforza-P. Menozzi-A. Piazza, *Storia e geografia dei geni umani*, Adelphi, Milan
Cifani **2008**: G. Cifani, *Architettura romana arcaica*, "L'Erma" di Bretschneider, Rome
Coarelli **1983:** F. Coarelli, *Il Foro romano - Periodo arcaico*, Quasar, Rome
Coarelli **1988**: F. Coarelli, *Il Foro Boario*, Quasar, Rome
Davidson **1985**: N. Davidson, *Astronomy and the Imagination - A new approach to man's experience of the stars*, Routledge & Kegan, London-New York
Della Corte **1969**: F. Della Corte, *L'antico calendario dei Romani*, Bozzi, Genoa
de Santillana-von Dechend **1983**: G. de Santillana-H. von Dechend, *Il mulino di Amleto*, Adelphi, Milan

Dumézil 1975: G. Dumézil, *Fetes romaines d'été et d'automne*, Gallimard, Paris
Dumézil 1977: G. Dumézil, *La religione romana arcaica*, Rizzoli, Milan
Dumézil 1981: G. Dumézil, *Mythe et épopée III*, Gallimard, Paris
Enea 1981:Enea nel Lazio - Archeologia e mito, Palombi, Rome 1981
Ernout-Meillet 1979: A. Ernout-A. Meillet, *Dictionnaire Etymologique de la Langue Latine*, Klincksieck, Paris
Etcheto 2012: H. Etcheto, *Les Scipions - Famille et pouvoir à Rome à l'époque républicaine*, Ausonius, Bordeaux
Fiorina 1985: P. Fiorina, *Etnoarcheologia*, in *La terra tra i due fiumi - Venti anni di archeologia italiana,* Il Quadrato Alessandria
Galileo 2009: *Galileo - immagini dell'universodall'antichità al telescopio*, (a cura di P. Galluzzi), Giunti, Firenze
Garin 1996: E. Garin, *Lo zodiaco della vita*, Laterza, Bari
Gli Etruschi, edited by M. Torelli, Palazzo Grassi, Venice, November 2000-April 2001, Bompiani, Milan
Gombrich 1978: R. F. Gombrich, *Antica cosmologia indiana* in *Antichecosmologie*, (edited by) C. Blacker-M. Loewe, Astrolabio, Rome
Graves 1992: R. Graves, *La dea bianca*, Adelphi, Milan
Hermann 1980: J. Hermann, *Atlante di astronomia*, Mondadori, Milan
Higgins 1997: R. Higgins, *Minoan and Mycenean Art*, Thames & Hudson, London
Invernizzi 1994: A. Invernizzi, *Il Calendario*, Quasar, Rome
Jacobs 1978: L. Jacobs, *Cosmologia ebraica*, in *Antiche Cosmologie*, edited by C. Blacker - M. Loewe, Astrolabio, Rome
Jones 1993: A. Jones, *Evidence for Babylonian Arithmetical Schemes in Greek Astronomy*, in H. D. Galter (edited by), *Die Rolle der Astronomie in den Kulturen Mesopotamiens*, Beitrage zum 3. Grazer Morgenlaendischen Symposion, 23–27 September 1991, Graz, pp. 77–94
Kazakhstan 2010: *Kazakhstan - homes, bêtes et dieux de la steppe*, Artlys, Versailles
Kent 1953: R.G. Kent, *Old Persian*, American Oriental Society, New Haven
Latte 1960: K. Latte, *Roemische Religiongeschichte*, Munchen
Les Etrusques 1992: Les Etrusques et l'Europe, Paris Reunion des Musées Nationaux
Liou-Gille 1980: B. Liou-Gille, *Cultes "heroiques" romains*, Les Belles Lettres, Paris
Lugli 1970: G. Lugli, *Itinerario di Roma antica*, Periodici Scientifici, Milan
Magini 1996: L. Magini, *Le feste di Venere - Fertilità femminile e configurazioni astrali nel calendario di Roma antica*, L'Erma di Bretschneider, Rome
Magini 2001: L. Magini, *Astronomy and Calendar in Ancient Rome - The Eclipse Festivals*, "L'Erma" di Bretschneider, Rome
Magini 2005: L. Magini, *Eclissi e regalità: un rapporto difficile (Babilonia 1900 a.C.-Roma 1600 d.C.)*, in "Rivista Italiana di Archeoastronomia", Vol. III, pp. 1–5
Magini 2006: L. Magini, *I moti di Venere e le feste delle donne nel calendario di Roma antica*, in "Rivista Italiana di Archeoastronomia", Vol. IV, pp. 123–34
Magini 2007: L. Magini, *L'etrusco, lingua dall'Oriente indoeuropeo*, "L'Erma" di Bretschneider, Rome
Magini 2008: L. Magini, *La dea bendata - lo sciamanesimo nell'antica Roma*, Diabasis, Reggio Emilia
Magini 2011: L. Magini *Controstoria degli etruschi - viaggio alle sorgenti orientali della civiltà romana*, "L'Erma" di Bretschneider, Rome
Marinone 1987: N. Marinone, *Introduzioneai* Saturnalia *di Macrobio*, Utet, Turin
Matres Matutae 1991: Matres Matutae dal Museo di Capua, Angelicum, Milan
Mayrhofer 1956: M. Mayrhofer *Kurzgefasstes etymologisches Woerterbuch des Altindischen*, Winter, Heidelberg
Meeus 1997: J. Meeus, *Mathematical Astronomy Morsels*, Willmann-Bell, Richmond Va.
Mountdorf 1923: J. F. Mountdorf, *De mensium nominibus*, in "Journal of Hellenic Studies", XLIII, pp. 102–16

North 2008: J. North, *Cosmos - An Illustrated History of Astronomy and Cosmology*, The University of Chicago Press

Nougayrol 1965 : J. Nougayrol, *Les rapports des haruspicines étrusque et assyro-babylonienne*, in "Comptes Rendus de l'Académie des Inscriptions et Belles-lettres", 1955, pp. 509 ff.

Nougayrol 1966: J. Nougayrol, *La divination en Mésopotamie ancienne*, in "RAI", XIV, Paris, 1966

Oppenheim 1974: A. L. Oppenheim, *A Babylonian Diviner's Manual*, in "Journal of Near Eastern Studies", 33, pp. 197–220

Pannekoek 1989: A. Pannekoek *A History of Astronomy*, Dover, New York

Pettinato 1998: G. Pettinato *La scrittura celeste*, Mondadori, Milan

Principietruschi 2000:*Principi etruschi tra Mediterraneo ed Europa*, edited by G. Bartoloni, F. Delpino, C. Morigi Govi, G. Sassatelli, Museo Civico Archeologico, Bologna October 2000-April 2001, Marsilio, Padua

Romano 1985: G. Romano, *Introduzione all'astronomia*, Muzzio, Padua

Romano 1994: G. Romano, *Archeoastronomia italiana*, CLEUP, Padua

Schiaparelli 1998: G. Schiaparelli, *Scritti sulla Storia dell'Astronomia antica, Tomo I-Tomo II-Tomo III*, Mimesis, Milan (anastatic reprint from Bologna, 1925)

Segni 2001: *Segni e sogni della terra*, De Agostini, Milan

Silvestri 1990: D. Silvestri - L. Tonelli - V. Valeri, *The Earliest Script of Uruk*, Ann. Dip. Studi Mondo Classico Mediterraneo Antico, Naples

Stumpff 1963: K. Stumpff (edited by), *Astronomia*, Feltrinelli, Milan

Swerdlow 1998: N. M. Swerdlow, *The Babylonian Theory of the Planets*, Princeton University Press

The Complete Ephemerides 1920–2020 (edited by F. Santoni), Aureas Editions, Paris 1995

The Gold 2008: *The Gold of Macedon - ArcheologicalMuseum of Thessaloniki*, Archaeological Publications Department, Athens

TLE 1968: M. Pallottino, *Testimonia Linguae Etruscae*, La Nuova Italia, Florence

Torelli 1984: M. Torelli, *Lavinio e Roma*, Quasar, Rome

vander Waerden 1949: B. L. van der Waerden, *Babilonian Astronomy II - The Thirty-Six Stars*, in "Journal of Near Eastern Studies", Vol. VIII, 1949, pp. 6–26

Walker 1997: C. Walker (edited by), *L'astronomia prima del telescopio,* Dedalo, Bari (original title: *Astronomy Before the Telescope,* British Museum Press, London, 1996)

Warde Fowler 1908: W. Warde Fowler, *The Roman Festivals of the Period of the Republic*, MacMillan, London

Yule 1981: P. Yule, *Early Cretan Seals: a Study of Chronology*, von Zabern

Index

L. Magini, *Stars, Myths and Rituals in Etruscan Rome*, Space and Society,
DOI 10.1007/978-3-319-07266-1

MIX
Papier aus verantwortungsvollen Quellen
Paper from responsible sources
FSC® C105338

If you have any concerns about our products,
you can contact us on
ProductSafety@springernature.com

In case Publisher is established outside the EU,
the EU authorized representative is:
Springer Nature Customer Service Center GmbH
Europaplatz 3, 69115 Heidelberg, Germany

Printed by Libri Plureos GmbH
in Hamburg, Germany